ELEMENTARY FOREST SURVEYING AND MAPPING

By
Robert L. Wilson
Associate Professor of Forest Engineering
School of Forestry
Oregon State University

Published by
O.S.U. Book Stores, Inc.
Corvallis, Oregon

1981

Litho--U.S.A.

PREFACE

Surveying done in Forestry whether it be a property line, timber cruising, or road location is usually done on adverse topography because that is where the timber grows, particularly in the Pacific Northwest. The principles of surveying are the same, but the approach is slightly different. It is in the best interest of the student if he receives his field training on rough terrain similar to that encountered on the job so that he may recognize the problems related to surveying in Forestry.

Elementary Forest Surveying and Mapping is designed as a text for an introductory course in Forest Engineering and attempts to cover the less expensive and less precise methods up to but not including transit and levels. Distance measurement by steel tape and pacing, direction by compass, and elevation differences by abney, clinometer, and aneroid barometer are emphasized. Much of the material in this text is not treated with sufficient detail in standard surveying text books.

A brief description of Public Land Survey is included in Chapter II to acquaint the student with the system.

Any operation on forest land involves the use and knowledge of topographic maps, whether it be the interpretation or the actual preparation and construction of such a map. Chapter VI describes one method of mapping in the field which involves the use of all the equipment and the principles and theories of elementary surveying as described in this text.

Slide rule instruction, abney tables, and trigonometric functions are included in the Appendix for use in solving basic problems encountered in surveying.

0-88246-135-4

CONTENTS

Page

PREFACE iii

CHAPTER I. INTRODUCTION 1

Section 1. General 1
definition of surveying—requirements of a good surveyor—less precise equipment.

Section 2. Field Notes 2
requirements of evaluation—common errors—numerical values—sketches—explanatory notes—abbreviations—sample note form.

CHAPTER II. PUBLIC LAND SURVEY 9

Section 1. Early Boundary Description 9

Section 2. History of the Development of the Public Land Survey System 10

Section 3. Provision of Public Land Survey Law 11
initial control points—meridians—base lines—townships—ranges—sections—subdivision of sections—description

Section 4. Subdivision of Township and Section 13
sections—descriptions

Section 5. Problems 21

CHAPTER III. MEASUREMENT OF HORIZONTAL DISTANCE 22

Section 1. Tapes 22
throwing—chaining—engineers tape—topographic tape—errors—corrections—noteform

Section 2. Pacing 30
percent of slope—methods of learning pace length—pacing tables—sources of error—auxiliary equipment

Section 3. Problems 34

CHAPTER IV. MEASUREMENT OF DIRECTION 37

Section 1. Angular Relationships 37
True bearing—azimuth—magnetic declination—magnetic bearing—local attraction—corrections for local attraction—deflection angles—interior angles

Page

Section 2. Compasses 46
hand compass–staff compass–declination correction

Section 3. Problems 52

CHAPTER V. MEASUREMENT OF VERTICAL DISTANCES 55

Section 1. Aneroid Barometer 55
description–principles of operations–errors–elevation corrections–field notes

Section 2. Clinometer 57
description

Section 3. Abneys 59
principles involved–types of adjustments–errors–slope tables for percent abneys–topographic tapes, abneys and trailers–fractional slope distances

Section 4. Leveling 68
equipment–definition of terms–note form–method

Section 5. Problems 71

CHAPTER VI. FOREST MAPPING 75

Section 1. Maps 75
description–uses–contours–scales

Section 2. Field Procedures 79
map control–horizontal and vertical–errors of closure–elevation corrections–strip mapping–detail and contour–location–work sample–strip closure

Section 3. Map Assembly 89
form–adjusting line and distance error in field work–strip detail–datum–legend

Section 4. Problems 93

APPENDIX 98

Slide Rule 99

Trigonometric Formulas 110

Table 1. Field table for percent abney. 113

Page

Table 2. Field table for compass and degree abney 114

Table 3. Conversion of degrees to percent and topographic graduations. 115

Table 4. Conversion of topographic graduations to percent and degrees. 117

Table 5. Conversion of percent to topographic graduations and degrees. 119

Table 6. Conversion of slope distance to horizontal distance for various readings on the topographic abney. 122

Table 7. Slope reduction table for the percent abney . . 123

Table 8. Natural trigonometric functions. 144

Problem Answers 169

Sample note forms 173

LIST OF ILLUSTRATIONS

Figure		Page
1.	Sample note form.	7
2.	Location of control points and meridians.	12
3.	Division by standard parallels and guide meridians.	14
4.	Subdivision in townships.	15
5.	Order of subdivision and numbering of sections in township.	17
6.	Subdivision of sections.	19
7.	Description of areas in section subdivision.	20
8.	Chaining (a) horizontal (b) slope (c) breaking chain.	24
9.	Engineer's tape (a) adding tape (b) subtracting tape Topographic tape (c) one chain mark and trailer (d) two chain trailer.	26
10.	Correction for slope chaining.	28
11.	Note form for chaining.	29
12.	True bearings.	37
13.	True bearings and azimuth.	38
14.	Isogonic chart of the United States, showing lines of equal magnetic declination.	40
15.	Relationship between true bearing and magnetic bearings.	41
16	True bearings and magnetic bearings.	42
17.	Corrections for change in magnetic declination.	42
18.	Traverse with deflection and interior angle.	44
19.	Local attraction in a traverse.	45
20.	Correcting for local attraction at station C.	46
21.	Correction for local attraction at station D.	46

Figure		Page

22. Box or hand compass 47

23. Silva hand compass. 48

24. Staff compass. 50

25. Pocket size aneroid barometer 56

26. Aneroid barometer note form. 57

27. Clinometer. 58

28. Eye view of clinometer. 59

29. Abney hand level. 60

30. Two peg method of adjusting abney hand level. 62

31. Measuring tree heights with percent abney 64

32. Slope chaining with topographic tape and topographic abney. 67

33. Hand level. 69

34. Level rod. 69

35. Differential level note form. 70

36. Contours (a) parallel planes intersecting terrain. 77
(b) sections cut by parallel planes. 77
(c) contour map of terrain in (a). 77

37. Contour map. 78

38. Map controls. 81

39. Locating contours in field. 83

40. Work sheet for strip mapping. 85

41. Map format. 90

42. Adapting field traverse to finished form. 91

43. Conventional signs and symbols. 95

ELEMENTARY FOREST

SURVEYING AND MAPPING

CHAPTER I. INTRODUCTION

Section 1. General

Surveying is the art of location of points or lines on or near the surface of the earth by measuring angles, directions, and distances. When a survey is of such a limited extent that the curvature of the earth can be neglected, it is called "Plane Surveying". If a survey is so large that the curvature of the earth must be taken into consideration, it is referred to as "Geodetic Surveying". Surveys are made for a variety of purposes such as (1) to determine the horizontal position and elevation of points on the earth's surface; (2) to determine the configuration of the ground; (3) direction and length of lines; (4) to determine the position of boundary lines and areas bounded by such lines, and (5) to determine the position of buildings, roads, dams, etc. These types of surveys may be better known as control surveys, boundary or property surveys, topographic surveys, and construction surveys.

A good surveyor must have not only technical knowledge and skill but he must also have good common sense and judgment and be willing to accept his responsibilities of honesty and accuracy. Traits of character are just as important or more so than technical knowledge and skill in making the reputation of a surveyor. An individual must also have or develop the ability to use good jugement, common sense, and ingenuity in all of the work he does. There is no place for error or mistakes in this type of work, and only correct judgment, answers, or solutions to problems are acceptable.

In surveying, one must decide if the project warrants the use of very accurate instruments which will result in higher costs of a project, or if less accurate instruments will suffice. For example, there is not much need for spending the time and money that would be required to make a transit location of a spur logging road if a compass location would be good enough considering the standards and specification to which the road would be constructed. The use which is to be made of the survey will help in making the decision concerning the cost and accuracy.

This text is mostly confined to the use and application of what are often called the less precise instruments; i.e., the tape and pace to measure horizontal distance, the compass to measure angles and direction, and the abney, clinometer, and aneroid barometer to measure the differences in elevation indirectly and directly. The steel tape is used for very precise measurements when such things as tension, support and temperature are measured and considered, but when the tape is used with auxiliary equipment such as the compass and abney, such precision is not needed because of the accuracy of the other instruments themselves.

Awareness must be maintained at all times against the attitude that the use of less precise equipment does not require accuracy. It is very easy to lapse into carelessness with this equipment. It must also be remembered that less precise work does not mean careless work. Measurements made to the nearest link or 1/10 of a foot and the nearest 1/4 of a degree will give satisfactory results if the surveyor has done consistent work throughout the survey. Errors produced are usually consistent and compensating. Most major errors are produced by personnel and not as a result of the equipment.

The advantage of these instruments is the simplicity and speed with which they are used. The student must learn the fundamentals first to avoid trouble and to concentrate on accuracy. Speed and ease will develop after sufficient practice and experience.

The process of surveying may be divided into field work and office work. The foregoing is only a part of the field work. The field notes are equally important.

Section 2. Field Notes

Field notes are permanent written records of surveys taken at the time the work was done in the field. They should be in such form and clarity that anyone else may readily interpret them. If the record of the field work is illegible or unreliable, the survey, no matter how carefully done, has been a complete waste of time, energy, and money. Five main features considered in evaluating field notes are ACCURACY, INTEGRITY, LEGIBILITY, ARRANGEMENT, and CLARITY.

Some of the common mistakes in note keeping can be eliminated by observing the following points.

a. Use a well-pointed 3H or 4H pencil. A piece of sandpaper taped in the back of the field book is handy to keep the pencil sharp.
b. Use the Reinhardt system of slope lettering for clarity and speed. Do not mix upper and lower case letters.
c. Make a neat title on the cover of the field book and on the first page inside the cover showing the owner or the company making the survey.
d. Leave a page or two in the front of the book, immediately following the title page for an index of the work done in the field. Keep the index up to date. The index should show the name of the problem, location, and the page number.
e. If the page of notes becomes illegible or if celluloid sheets are used for recording notes in wet weather, make a copy of the data while the information is still fresh in mind. Then mark it COPY in the field note book.
f. If a page is to be voided, draw diagonal lines from the opposite corners and letter VOID prominently but do not obscure any numerical values or any part of the sketch.
g. The left hand page contains the numerical values for the right hand page, and the two pages are practically always used in pairs. Therefore, they carry the same page number, which should be placed in the upper left hand corner of the left page and the upper right hand corner of the right page.
h. Always record directly in the field book rather than on a scrap of paper for copying later.

The forms and methods of notekeeping are different for various types of survey work. Information recorded in the field notes are generally classified into three parts: (1) numerical values, (2) sketches, and (3) explanatory notes.

Numerical values are tabulated records for all measurements and are recorded from the bottom of the page toward the top, with the exception of level notes. Specific instructions regarding these are as follows:

a. Make large plain figures.
b. Never write one figure on top of another.
c. Avoid trying to change one figure into another.
d. Erasing is prohibited. Draw a line through the incorrect value and write the correct value directly above.
e. Repeat aloud values for recording. For example, before recording 143.57 call out one, four, three, point, five, seven for verification.
f. Place a zero before the decimal point for numbers less than one and show the precision of measurements by recording significant zeros.
g. Do not use fractions in recording compass bearings which are not even degrees.

Sketches are graphic records of boundary outlines, relative locations, topographic features, or any diagram to further clarify the tabulated values.

a. They are rarely drawn to scale. It may be necessary to exaggerate certain portions of the sketch for purposes of clarity.
b. Use a straight edge for sketches (6 inch celluloid protractor).
c. The red line in the center of the right page may be used to represent the route of travel on a traverse.
d. Make sketches large, open, and clear.
e. Line up description and sketches with corresponding numerical data if possible.
f. Avoid crowding.
g. When in doubt about the need for recording any information, include it in the sketch.

Explanatory notes clarify the numerical values and sketches that might otherwise be misunderstood. Such notes are always printed. In all cases one should ask himself if the numerical values and the sketches require additional explanation for easy interpretation. The following explanatory notes are usually included in every set of field notes and are recorded on the right hand page with the exception of the title.

a. Title of project or problem at top of left hand page.
b. Location of project at top of right hand page.
c. Date of project.
d. Names of the survey party.
e. Duties of each member of the survey crew.
f. Equipment used, with identifying numbers, if any.
g. Weather conditions.
h. Show true north, magnetic north, and magnetic declination.

Some of the more common abbreviations and symbols used in field notes are listed as follows:

Sta.	station	Mg. Brg.	magnetic bearing
H.D.	horizontal distance	Elev.	elevation
S.D.	slope distance	Int. ∡	interior angle
D.E.	difference in elevation	Calc.Brg.	calculated bearing
T.N.	true north	Mg. Decl.	magnetic declination
M.N.	magnetic north	F.S.	foresight
∡	angle	B.S.	backsight
Defl. ∡	deflection angle	Comp.	compass
Vert. ∡	vertical angle	H.C.	head chainman
Az.	azimuth	R.C.	rear chainman
Brg.	bearing	C.C.	crew chief

C.P.	chief of party	Sec. Cor.	section corner
B.M.	bench mark	¼ Sec. Cor.	¼ section corner
T.P.	turning point	W.M.	Willamette Meridian
△	triangulation station	℄	center line
H.I.	height or elevation of instrument	H. ⚑	head flagman
T.	township	R. ⚑	rear flagman
R.	range	⊙	transit station
Sec.	section	⊼	transit or level

An example of one method of keeping notes in the field book is illustrated in Figure 1. Notice that one of the stations is located in reference to a known point; hence anyone following the recorded distances and directions could retrace each line. Usually the notes and sketch are found on corresponding lines with the exception of a closed traverse, but the sketch is still oriented with north toward the top of the page.

4/

5 SIDED TRAVERSE

STA H.D	F.S.	B.S.	INT ∢	CORRECT INT ∢	CALC. BRG
B					
458.31	N32°30'W	S33°30'E			N32°30'W
E					
435.83	N1°00'W	S0°30'E			N1°00'W
C					
A			124°30'	124°30'	
294.20	N62°00'W	S62°00'E			N62°00'W
E			95°00'	95°00'	
342.61	S31°00'W	N33°00'E			S33°00'W
D			101°00'	101°30'	
285.65	S45°00'E	N48°00'W			S45°30'E
C			125°00'	125°00'	
292.50	N79°30'E	S79°00'W			N79°30'E
B			93°30'	94°00'	
310.09	N6°30'W	S7°00'E			N6°30'W
A					
		Σ	539°00'	540°00'	

Figure 1. Sample note form.

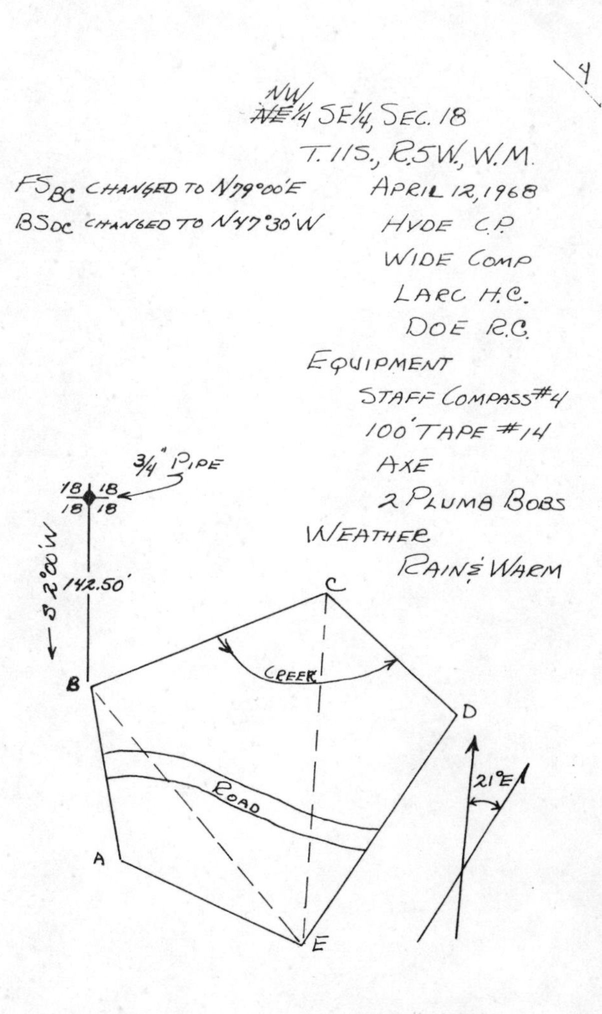

Figure 1. (cont) Sample note form.

CHAPTER II. PUBLIC LAND SURVEY

Section 1. Early Boundary Description

Nearly all of the land grants in the colonies were irregularly shaped, the boundaries following shore lines, streams, fences, or ridges. There was no system to serve as a control for the boundaries, and it was almost impossible to restore them if they became obliterated. This type of surveying was known as Rural Land Description, and it preceded the system of "Metes & Bounds" which was done by giving the length and direction of the boundaries of a piece of land. As the country developed and the land became more valuable, land litigations became more numerous. The following description, taken from the court records in 1812 in the State of Connecticut illustrates what is encountered in Rural Land description.

"147 acres, 3 rods, and 19 rods after deducting whatever swamp, water, rock and road areas there may be included therein and all other lands of little or no value, the same being part of said deceased's 1280 acre colony grant, and the portion hereby set off being known as near to and on the other side of Black Oak Ridge, bounded and described more in particular as follows, to wit:—Commencing at a heap of stone, about a stone's throw from a certain small clump of alders, near a brook running down off from a rather high part of said ridge; thence, by a straight line to a certain marked white birch tree, about two or three times as far from a jog in a fence going around a ledge nearby, thence, by another straight line in a different direction, around said ledge and the Great Swamp, so called; thence, in line of said lot in part and in part by another piece of fence which joins on to said line, and by an extension of the general run of said fence to a heap of stone near a surface rock; thence, as aforesaid, to the 'Horn' so called, and passing around the same as aforesaid, as far as the 'Great Bend', so called, and from thence to a squarish sort of a jog in another fence, and so on to a marked black oak tree with stones piled around it; thence by another straight line in about a contrary direction and somewhere about parallel with the line around by the ledge and the Great Swamp, to a stake and stone bounds not far off from the old Indian trail; thence, by another straight line on a course diagonally parallel, or nearly so, with 'Fox Hollow Run', so called, to a certain marked red cedar tree out on a sandy sort of a plain; thence, by another straight line, in a different direction, to a certain marked yellow oak tree on the off side of a knoll with a flat stone laid against it; thence, after turning around in another direction, and by a sloping straight line to a certain heap of stone which is by pacing, just 18 rods and about one half a rod more from the stump of the big hemlock tree where Philo Blake killed the bear; thence, to the corner begun at by two straight lines of about equal length, which are to be run by some skilled and competent surveyor, so as to include the area and acreage herein before set forth."

Section 2. History of the Development of the Public Land Survey System

Our present day pattern of Land Survey owes its rectilinear form to the Jefferson-Williamson plan of land subdivision proposed in 1784. Jefferson's proposal was based on a geographical mile, equivalent in the 18th Century to the nautical mile which is 6,086.4 feet. This was obtained by taking the currently accepted value of a degree of latitude and dividing by sixty. Jefferson proposed to divide the geometrical mile into furlongs, each of these into 10 chains and each of these into 10 paces, differing very little from the British furlong, chain, and fathom.

A Jeffersonian Hundred was to be a tract of land 10 geographical miles on a side and containing 100 lots, each 100 reformed chains on a side. Since the surveying was to be done with a circumferentor, an instrument similar to our staff compass, it was necessary to note the magnetic variation. True bearings could be obtained by observing Polaris or observing the sun at the time of its rising or setting. Jefferson favored the later method and recommended that a copy of the table of amplitudes be furnished every surveyor. By amplitudes he meant the angular distance north or south of due east at which the sun would rise at a given date.

In 1796 the Federal Congress enacted the basic law under which, with few changes, practically all of the land north of the Ohio River and west of the Mississippi River has been surveyed. The land in Alabama, Mississippi, and Florida has also been surveyed by this system.

Section 3. Provision of Public Land Survey Law

The survey system begins with the establishment of an initial control point, whose longitude and latitude were determined by astronomical observations. The law provided for the establishment of 35 such points, and all land surveyed in an area is referred to one of these points. In Oregon and Washington this is known as the Willamette Meridian. This point is located south and west of Portland, Oregon. Figure 2 shows the location of these points.

A principal meridian is established as a true meridian through each initial point either north or south or both directions as conditions require. Permanent quarter-section and section corners were established alternately at intervals of 40 chains (½ mile) and township corners were placed at intervals of 480 chains (6 miles).

A base line through the initial point was extended east and west on a true parallel of latitude. Quarter section corners and section corners were established alternately at intervals of 40 chains (½ mile) and standard township corners at intervals of 480 chains (6 miles).

Figure 2 shows the name and location of the 35 initial control points.

The next step in the subdivision of the land being surveyed was to establish standard parallels. They were established in the same manner as the base line and are located at intervals of 24 miles north and south of the base line.

Guide meridians were established, and they divided the area into tracts which were approximately 24 miles square. These lines are true meridians which start at points on the base line on the standard parallels at intervals of 24 miles east and west of the principal meridian and extend north to their intersection with the next standard parallel. Because of convergence of the meridians, the distance between these lines will be 24 miles only at the starting point. Monuments are located at half mile intervals on all standard parallels and guide meridians.

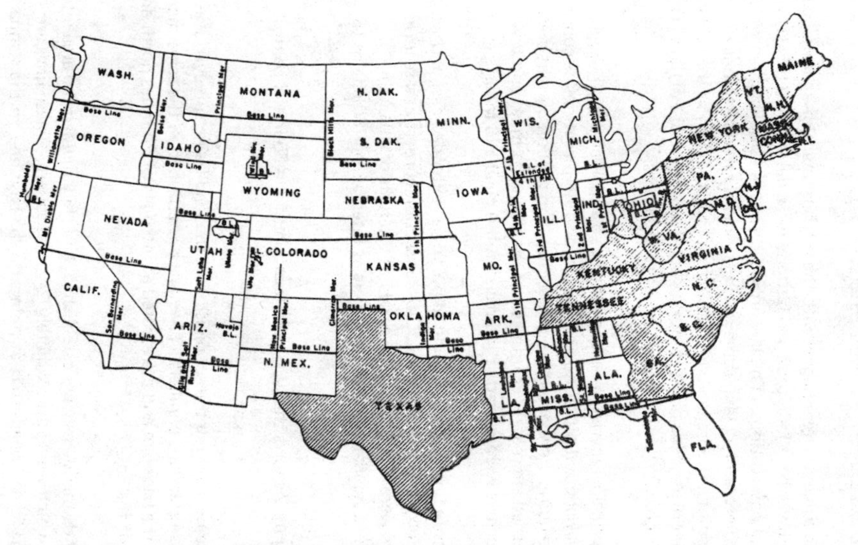

Figure 2. Location of control points and meridians.

Figure 3 shows the division of an area into tracts approximately 24 miles on a side by the Guide Meridians and Standard Parallels.

The next provision of the law was to divide the 24 mile quadrangles into townships by laying off range lines, at intervals of 6 miles along each standard parallel, which extend north 24 miles to the next standard parallel. The township corners established at intervals of 6 miles on the range lines, guide meridians, and principal meridians were joined with lines known as township lines. A row of townships extending north and south is called a range, and a row extending east and west is called a tier.

Both the east and west boundaries of all townships will be 6 miles in length but the north and south boundaries will vary in length from a maximum at the south parallel or base line to a minimum at that forming its northern boundary due to convergence of the range lines. All townships contain less than 36 square miles.

Figure 4 shows the division of a 24 mile tract into ranges and townships.

Section 4. Subdivision of Township and Section

The last step in the public land survey system was the division of the township into sections, each approximately 1 mile square and containing approximately 640 acres. This subdivision was done by establishing section lines, parallel to the east boundary, at intervals of 1 mile along the southern boundary. The purpose of this was to throw the irregularities into the north and west tiers of sections in each township. The sections are numbered from 1 to 36 beginning in the northeast corner of the section. The subdivision began at the corner of section 35 and 36 on the south boundary of the township and the line between sections 35 and 36 was run north parallel to the east boundary of the township. The quarter section corner was established at 40 chains, and the section corner common to section 25, 26, 35, and 36 was established. From this corner, a random line was run east, parallel to the south boundary of the township, to its intersection with the east boundary of the township. A temporary quarter corner was set on the random line running east and set permanently at the point midway

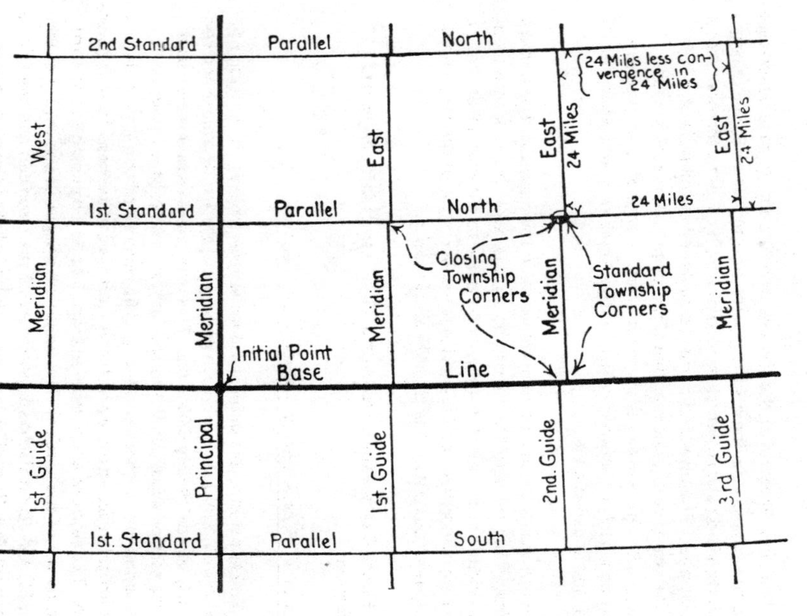

Figure 3. Division by standard parallels and guide meridians.

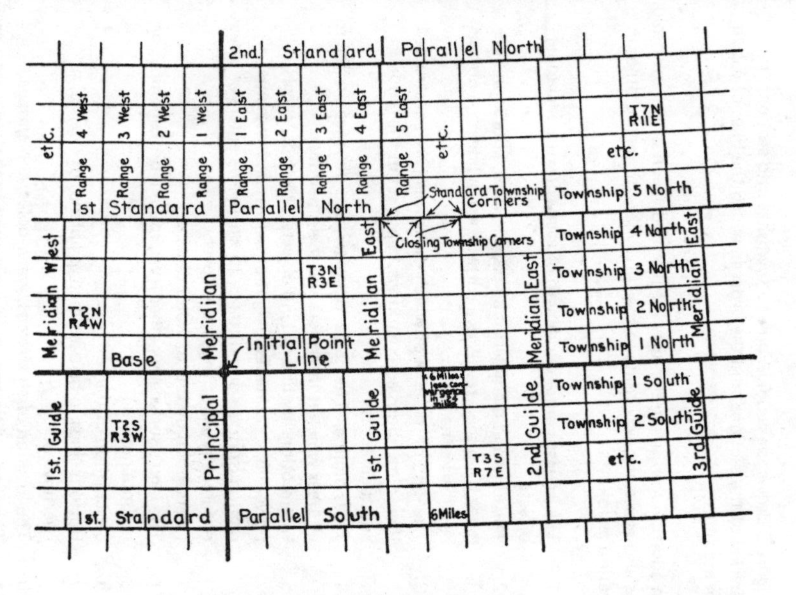

Figure 4. Subdivision in townships.

between the two section corners as the line was rerun from east to west. The line was then run north between sections 25 and 26.

Figure 5 shows the subdivision of a township into sections, the numbering of the sections, and the order in which the lines of subdivision are run.

In addition to running the lines for the subdivision of the township, it was also necessary, according to the survey law, to record other information as follows:

1. The precise course and length of every line run including offsets.
2. The kind and diameter of all bearing trees, course and distance from their respective corners, and all bearing objects and marks thereon.
3. The kind of material of which the corners are constructed, their dimensions and markings, depth set in the ground, and their accessories.
4. Trees on line. The name, diameter, and distance on line to all trees which it intersects and their markings.
5. Intersections by line of land objects, railroads, canals, power lines, estimated height of ascents and descents, direction of ridges and distance to where line enters or leaves different types of vegetative cover.
6. Intersection by line of water objects.
7. The land's surface, whether level, rolling, broken, hilly, or mountainous.
8. Description of the soil.
9. Bottom lands to be described as upland, or swamp, or overflowed.
10. The location of all streams, water-holes, and springs.
11. The kinds of timber and undergrowth in the order they predominate.
12. Lakes and ponds and whether the water is pure or stagnant, deep or shallow.
13. Improvements such as towns and villages, houses, field improvement, and official monuments not belonging to the survey system.

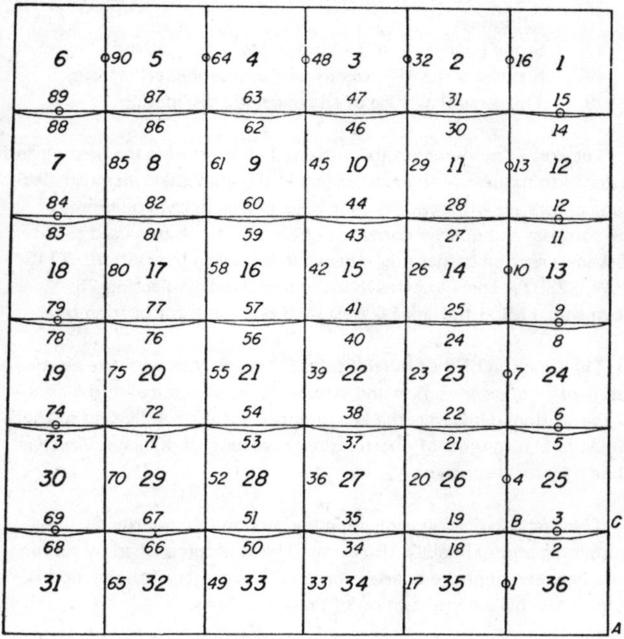

6 90 5 64 4 48 3 32 2 16 1
89 87 63 47 31 15
88 86 62 46 30 14
7 85 8 61 9 45 10 29 11 13 12
84 82 60 44 28 12
83 81 59 43 27 11
18 80 17 58 16 42 15 26 14 10 13
79 77 57 41 25 9
78 76 56 40 24 8
19 75 20 55 21 39 22 23 23 7 24
74 72 54 38 22 6
73 71 53 37 21 5
30 70 29 52 28 36 27 20 26 4 25
69 67 51 35 19 B 3 C
68 66 50 34 18 2
31 65 32 49 33 33 34 17 35 1 36
A

Figure 5. Order of subdivision and numbering of sections in township.

14. Coal beds, ore deposits, mining operations, and salt licks.
15. Direction of all roads and trails.
16. Rapids, cataracts, or water falls, their position and estimated fall in feet.
17. Stone quarries and the kind of stone they afford.
18. Natural curiosities, fossils, and archaeological remains.
19. The general average of the magnetic declination.

The corner monuments are witnessed by measuring the bearing and distance to natural or artifical objects in the immediate vicinity. Section corners are witnessed by four objects, one in each quadrant of the compass and quarter corners by two objects. A tree used as witness for a section corner might bear the following inscription: T11S, R5W, S23BT. This indicates that the tree stands in Section 23, Township 11 South, Range 5 West, and BT stands for bearing tree.

The function of the United States Surveyor ends with the establishment of quarter section and section corners on the external lines of the section. However, the law provided for the subdivision of the public land into units of quarter-quarter section of 40 acres, which is done by a local surveyor.

The first subdivision is dividing the section into quarter sections containing approximately 160 acres. This is accomplished by running lines between opposite quarter corners. The center of the section is determined by the intersection of these two lines.

The method of dividing these quarter sections into 40 acre tracts depends on the position of the section within the township. For any section except those along the north and west side of the township, the subdivision is accomplished by bisecting each side of the quarter section and connection of the points by straight lines. The intersection of these straight lines is the center of the quarter section. For sections along the north and west side of the township the subdivision of the quarter section is made in such a manner that any discrepancies from 20 chains is placed in the last quarter mile along the north side of the township and along the west side of the township (see Figure 6).

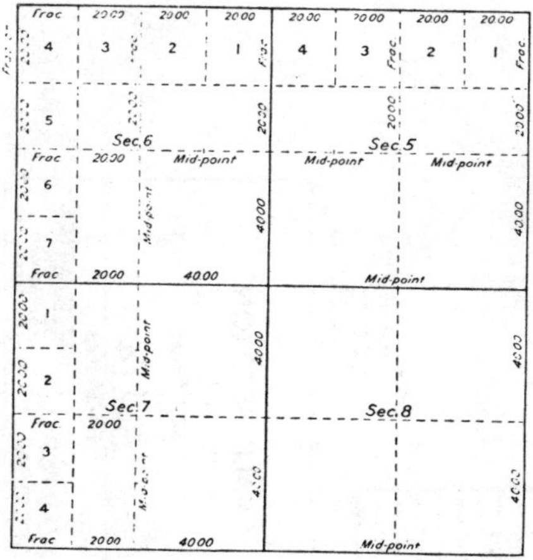

Figure 6. Subdivision of sections.

This system of land survey provides a very convenient method of describing a piece of land. In the description of a 10 acre tract, the quarter of the 40 acre tract is given first, followed by the quarter of the quarter section, then the quarter section, the section, township, range, and principal meridian.

Figure 7 shows the division of a section into smaller tracts.

The best reference for further information is the "Manual of Instructions for the Survey of the Public Lands of the United States" published by the government printing office. Circular #20 entitled "Digest of Oregon Surveying Law" available from the Engineering Experiment Station at Oregon State University is valuable for those surveying in Oregon.

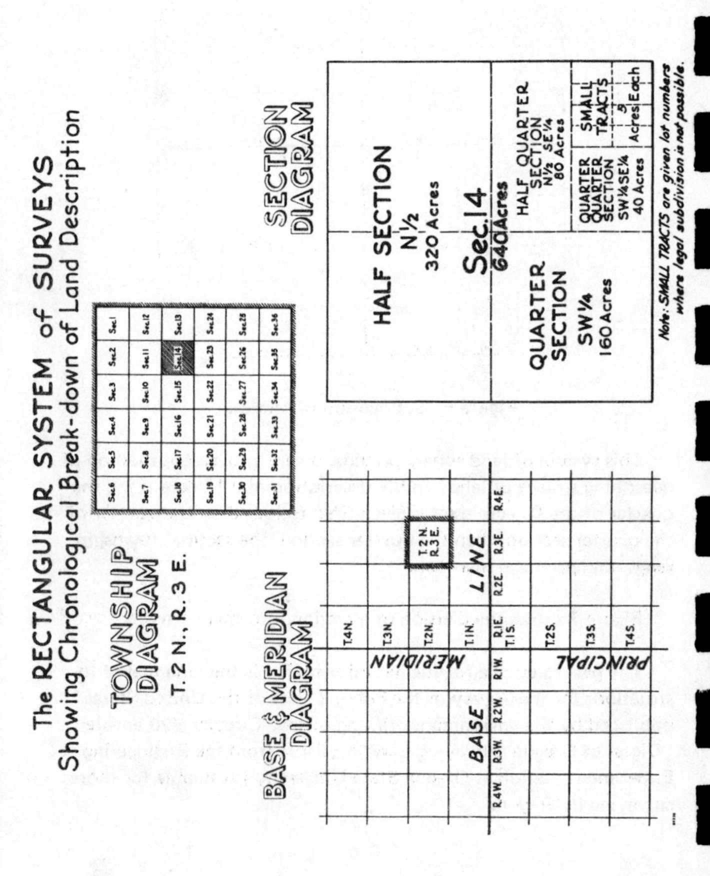

Figure 7. Description of areas in section subdivision.

Section 5. Problems

1. Approximately how many acres are in each of the following tracts?
   NE¼NW¼, Sec. 16
   N½NE¼, Sec. 16
   SE¼NW¼SW¼, Sec. 16
   E½SE¼, Sec. 16

2. How far is the NE section corner of Sec. 23, T.11S., R.5W., W.M. from the SW section corner of Sec. 13, T.11S., R.5W., W.M.?

3. In what order would the following corners normally be located?
   N¼ corner Sec. 36, T.11S., R.5W., W.M.
   NW section corner, Sec. 25, T.11S., R.5W., W.M.
   NE section corner, Sec. 35, T.11S., R.5W., W.M.
   S¼ corner, Sec. 26, T.11S., R.5W., W.M.

4. Show in a sketch by cross hatching the following tracts located in Sec. 9.
   a. SW¼NE¼, Sec. 9
   b. NW¼, Sec. 9
   c. SE¼NE¼SW¼, Sec. 9
   d. W½SW¼, Sec. 9

5. How would you differentiate between a line tree and a bearing tree?

6. The corner being considered is the southwest section corner of Section 31, T.6N., R.5W., W.M. What is the legal description of each "40" which share this common corner?

CHAPTER III. MEASUREMENT OF HORIZONTAL DISTANCE

Section 1. Tapes

Prior to taking a tape into the field the student should master the technique of "throwing" the tape, i.e., converting it from the figure eight in which it was coiled to the compact circle for ease in carrying, and he should also learn how to coil and uncoil the tape. These two operations are most easily mastered by actual practice.

There are several ways to coil a tape. One practical method is as follows: Take the zero end, tabs facing up, in the left hand, which is held palm up, and turn your back to the tape. The tape is then strung out behind past your right side. With the right hand, fingers under the tape and pointing toward the body, reach back along the tape a good arm's length (usually five feet for the engineers tape and eight links for the topographic tape) at the same time moving the left hand ahead and without turning the tape over, bring that tab up and lay it face up on the first tab. This is repeated until the entire length is coiled. The coils should all be of equal length. The tape is uncoiled with the same technique, laying off instead of on so that the top coil is taken off away from you. With practice, one need not look at the tabs to be sure of the distance each time. He becomes familiar with his reach and will be able to look and walk ahead while coiling or uncoiling the tape.

The steel tape is easily broken when treated carelessly. The chief danger is pulling on it when it is looped or kinked. Avoid jerking the tape, stepping on it, allowing vehicles to pass over it, or bending it around sharp corners. Wipe it dry before putting away. If it is to be stored for any length of time, other than overnight, spread a light coat of oil over it. Do not use the leather thongs at the end of the tape to tie it up if the thongs are wet. The steel tape when treated with care will last a lifetime.

Note that the tape was coiled starting with the zero end. Likewise, when it is uncoiled by the head chainman, he will finish with the zero end in his hand. The head chainman always stays ahead with the zero end of the tape. He is the one charged with the reading of fractional units off the graduated foot. The engineers tape may, however, be coiled beginning at either end providing the entire coil is turned inside out if need be so that the head chainman takes the zero end ahead. The topographic tape on the other hand, must be coiled beginning at the zero end because the trailer is not calibrated uniformly to permit the formation of even coils.

The horizontal distance between two points may be measured directly by keeping the tape horizontal or by measuring along the sloping ground and applying a correction to the measured distances or by reducing the slope distance to horizontal distance by trigonometry. In either case the head chainman walks ahead along the line with the zero end of the tape. The rear chainman stops him with a command of "chain". The rear chainman then lines up the head chainman on line and checks the vertical alignment of the tape either by eye or with the aid of a hand level if he is inexperienced. This is done only if the horizontal distance is to be measured directly rather than measuring the slope distance and computing the horizontal distance.

When the horizontal and vertical alignment are checked by the rear chainman, he calls out "stick" or "mark". The head chainman applies tension to the tape and marks his point on the ground by using a plumb bob or an axe. If necessary, the measurement is rechecked before recording in the field book; then the crew moves ahead with the head chainman pulling the tape. In forest surveying, the head chainman carries an axe to brush out the line, and he cuts his own pins or stakes to mark each measurement. Depending upon the terrain, the survey crew must decide whether to measure horizontal distance in full tape length or break chain by measuring short distances less than a tape length but still keeping the tape horizontal, or to measure slope distance and then reduce to horizontal distance by trigonometry (see Figure 8).

When the last measurement on a line is made, the distance usually is a fractional part of a tape length. With the engineers tape, the rear chainman will hold an even foot mark on the last point marked, and the head chainman will determine the fractional part of a foot by moving the plumb bob string along the tape until the plumb bob is over the point. The plumb bob string is placed over the tape and held in position with the thumb on top of the tape and string, and the index finger placed under the tape.

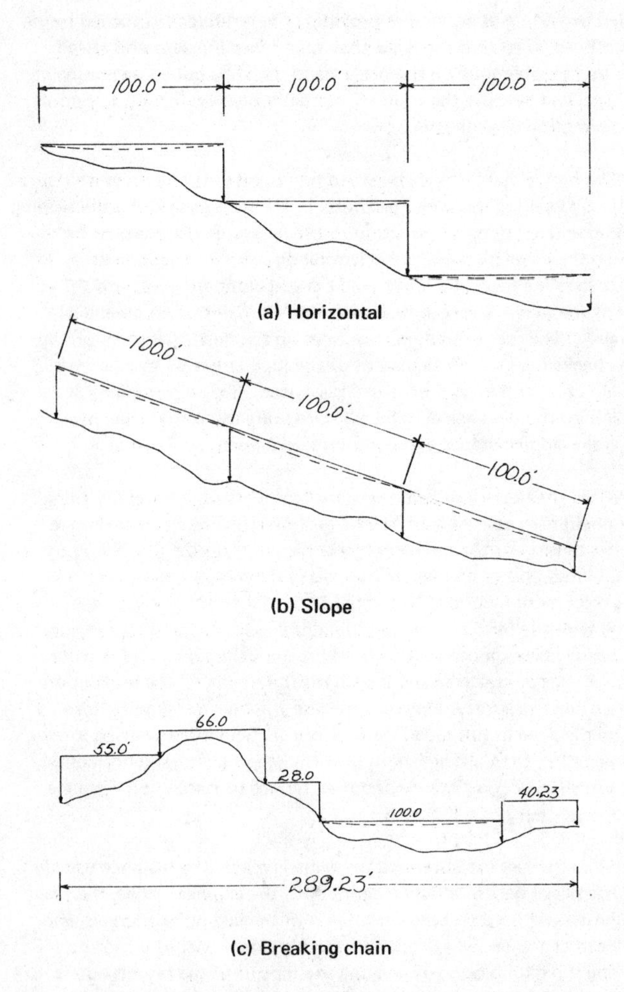

Figure 8. Chaining.

If the topographic tape is being used, the head chainman holds the zero mark on the end point, and the rear chainman reads the distance to the nearest half link over the last point set on the line.

Engineer tapes are made in 100, 200, and 300 foot lengths. These tapes are made of metal, but there are also tapes made of cloth called "rag tapes" which are satisfactory for setting slope stakes. The engineers tapes are graduated in even feet with the first or last foot, and sometimes both, graduated into 1/10ths or 1/100ths of feet in order to measure fractional distances. Engineer tapes vary in thickness and width depending on the length. Engineer tapes are also divided into two groups known as adding tapes or subtracting tapes depending on the method of graduations of the first foot. In measuring with an adding tape, the fractional part of the foot is added on to the foot value held by the rear chainman. However, with the subtracting tape, the fractional part of a foot held by the head chainman is subtracted from the value held by the rear chainman (see Figure 9a, b). The adding tape is the most desirable and is less subject to error. The chainman must examine each tape carefully before using to determine the type. The subtracting tape is always graduated in fractional parts of a foot between the zero mark and the one-foot mark.

Measuring a distance with a steel tape is commonly known as chaining. This term is derived from the early days of surveying when a Gunter Chain was used. This chain was 66 feet long and contained 100 steel-wire links, each 0.66 feet or 7.92 inches long. Hence, the word chain. There is always confusion regarding certain terminology and double meaning of certain words. To clarify this, remember the term "chain" is a unit of measurement of 66 feet. Taping or chaining is the act of measuring, and a tape is the instrument with which the distance is measured.

A Gunter chain is now called a surveyors tape, and it is a steel tape usually 2 chains or 132 feet long plus a trailer for the 2 chain tape. The tape is divided into link units which are 1/100 of a chain or 0.66 feet. The link and chain unit has a convenient relationship to the public land survey system in that there are 5 chains in a tally and 16 tallies or 80 chains per mile and 10 square chains per acre. The chain is also divided into quarters, a distance of 25 links, which equals 1 pole, or 16½ feet, or 1 rod. These last terms have become almost obsolete. Distances measured to the nearest ½ link when using a surveyors tape are usually accurate enough.

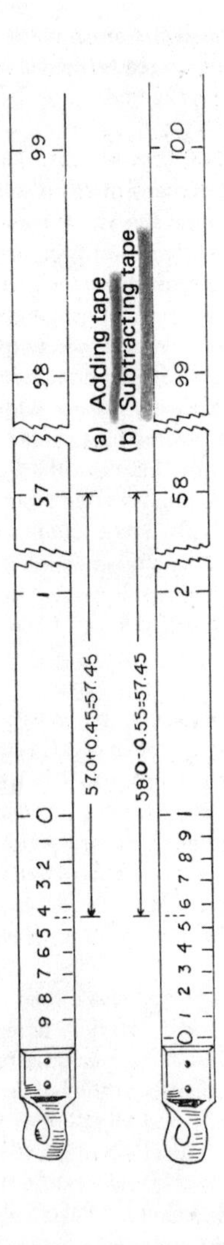

Engineer's tapes

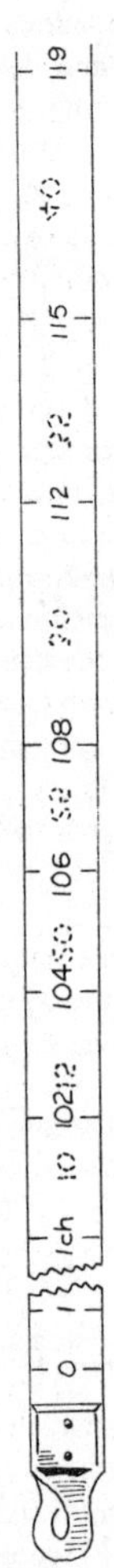

(c) One chain mark and trailer

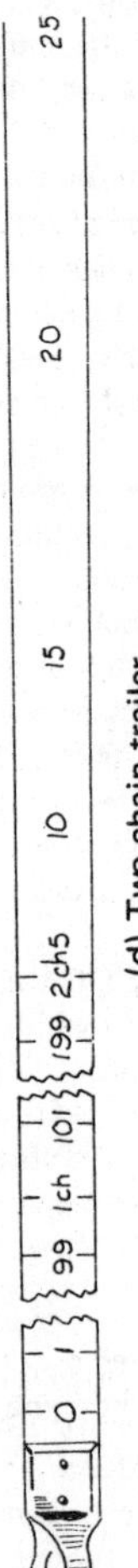

(d) Two chain trailer

Topographic tape

Figure 9. Tapes

The principal sources of error in linear measurements made with tapes are due to (1) incorrect length of tape (2) tape not horizontal (3) temperature variation (4) tension not standard (5) sag (6) improper alignment (7) tape not straight and (8) mistakes and errors.

The surveyor has control over the measurements made in the field if the tape being used is not of standard length. Most tapes are not of standard length because they may be slightly stretched or mended or have kinks in them. The distance measured with non-standard length tapes must be corrected accordingly. A distance measured with a long tape will be too short, and a short tape will give a measured distance greater than the true distance. It should also be remembered that just the opposite is true if a given line is to be laid out with a non-standard tape, i.e., when using a tape known to be short, the measured distance must be greater than the desired distance.

To correct for non-standard tapes, the following formula is used,

$L = \frac{TM}{S}$ Let S equal the standard length of the (100′, 200′, 300′ or 1 or 2 chain) tape.

T equals true length of tape

L equals the true length of the line.

M equals the measured length of the line.

Then T/S is the length of the tape in terms of standard. Knowing 3 of the unknowns, the 4th can be readily found. Thus the following equations could be used.

$$T = \frac{SL}{M} \qquad S = \frac{TM}{L} \qquad M = \frac{SL}{T}$$

The survey crew, particularly the rear chainman, is responsible for alignment both vertical and horizontal and seeing that the tape is not deflected out of line by trees or brush. Mistakes and errors of adding or dropping a full tape length, adding or dropping a foot, taking the wrong point for zero or 100 foot mark and reading wrong numbers (89 for 68), etc., are responsibilities of both head and rear chainmen. If one end of a hundred foot tape is held 1.41 feet higher or lower than the other end, the error will be 0.01 feet. While this error is not great, it can be cumulative and must be considered when chaining over rough terrain. The same amount of error occurs in alignment (Figure 10).

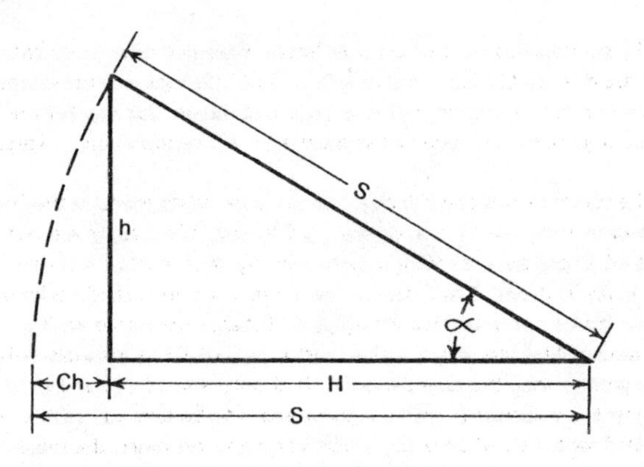

Figure 10. Correction for slope chaining.

S = slope distance
h = difference in elevation (S x sin∝)
H = horizontal distance (S x cos∝)
C_h = correction or (S - H) (S x vers∝)

$$S^2 = h^2 + (S - C_h)^2; S^2 - h^2 = (S - C_h)^2 \text{ and } C_h = S - (S^2 - h^2)^{1/2}$$

$$\text{by expansion } C_h = \frac{h^2}{2S} + \frac{h^4}{8S^{3/2}} \quad \text{(approx)}$$

For most measurements made in Forest surveying, temperature is not considered, but it should be noted that a change of 15 degrees in temperature will cause a change of approximately 0.01 feet in a 100 foot tape.

Too much tension causes a tape to stretch, but not enough tension results in too much sag. An increase of 10 lbs. in tension will stretch a light 100 foot tape about 0.01 feet and a heavy tape about 0.003 feet. A sag of approximately 8" in the center of a 100 foot tape will shorten the measured distance by 0.01 feet. Sag may be compensated for by increasing the tension from 10 to 18 lbs. for a light tape and to 50 lbs. for heavier tapes. While a pull of 50 lbs. is difficult, it is not impossible.

The recording of measurements made in the field is shown in Figure 11. The horizontal distances measured between the stations are recorded on the lines between the stations. Stations, which are defined as the accumulated distance from the beginning point, are shown in one of three methods, depending upon the type of survey and tape being used. A plus sign (+) is used to note the stations when the measurements are in feet but is never used in the distance itself. The decimal (.) is used to note the stations when the measurements are in chains and links. Letters may also be used in small closed traverses where the accumulated distances are not needed.

In Figure 11, it should be noted that the data is recorded from the bottom to the top of the page. In this manner the sketch will conform with the data on the left page. See note forms in appendix.

Sta	H.D.	Sta	H.D.	Sta	H.D.
$5+49^{\underline{6}}$		5.07		E	
	200.0		2.00		106.8
$3+49^{\underline{6}}$		3.07		D	
	61.4		0.82		50.7
$2+78^{\underline{2}}$		2.25		C	
	152.8		1.25		193.6
$1+25^{\underline{4}}$		1.00		B	
	125.4		1.00		149.4
$0+00^{\underline{0}}$		0.00		A	
Engineers Tape		Topographic Tape		Lettered Stations	

Figure 11. Note form for chaining.

Section 2. Pacing

A pace is a measurement of horizontal distance. Regardless of the slope it is always recorded in terms of horizontal distance. It is most handily used in retracing lines for locating public land subdivision corners. Timber cruisers use the pace extensively, and for the construction of an inexpensive less precise type of map the pace used with the hand compass is often quite satisfactory.

Since no one but the individual concerned knows the length of his respective pace, paces as such are never recorded in the field notebook as a unit of measurement. They are always converted to some standard unit, such as feet or chains, before recording in the notebook.

It is advisable at this point to define percent of slope, a term to be used frequently hereafter. It is common practice to give the grade of a slope in terms of percent. This is not to be confused with the "vertical angle," the angle between the horizontal plane and the line of sight along a slope. Percent of slope is the number of feet rise or fall vertically per 100 feet on the horizontal. For example, a difference of one foot vertically in a hundred feet of horizontal distance is equal to a plus or minus one percent depending on whether the difference in elevation is plus or minus. A rise of 10 feet vertically in 200 feet on the horizontal will equal 5 percent. By formula:

$$\% \text{ slope} = \frac{\text{vertical distance} \times 100}{\text{horizontal distance}} = \frac{10 \times 100}{200} = 5\%$$

In learning to pace, one must become proficient in allowing for slope, for the steeper the slope, the more extra steps one must take to get the desired horizontal distance. One method is to set up a table of paces showing the number of paces an individual takes on level ground and on various slopes. This method has its limitations because the slope is seldom uniform.

Courses are laid out a distance of 1 tally each (5 chains or 330 feet). Each course is traversed five times, and an average is taken of the number of paces. Any count differing from the others by three paces or more is thrown out, and the course is traversed again.

When the table is completed, the pacer has his average number of paces per tally on the level. In addition to this basic information, he knows how many more paces he takes on various slopes than he does on the level. He is also able to compute the length of his double pace (it is optional whether the pacer uses double or single paces, but the double pace is usually preferred). To make the table more useful, he should convert his paces per tally to paces per chain and to paces per hundred feet.

When this table has been completed, it will be noted that as the percent of slope increases the number of paces also increases. The horizontal distance of each course remains the same, but the slope distance has increased. Also the length of the pace changes with the slope.

From the pacing table, one might find that the number of paces taken on the various slopes as compared to the number of paces on the level to be as follows:

Slope %	Number of extra paces up	Number of extra paces down
10	½	0
20	2	1
30	3½	2
40	5	3

The Forest Service has developed a method by which the number of paces per chain or tally on any slope is the same as the number of paces for the same distance on the level. This is accomplished by skipping or not counting a pace at certain intervals, depending on the slope. This method requires the pacer to possess the ability to estimate slope. A sample table of pacing by this method, once the number of paces on 0% slope is known, would be as follows:

%	Up		Down	
Slope	Paces	Skip	Paces	Skip
10	6	1	—	—
20	3	1	11	1
30	2	1	6	1
40	1	1	2	1

A third method, which can be easily adapted by the beginner, is to lay out 4 or 5 courses on typical terrain through the brush, each course one tally in length. The individual then paces up and down each course without differentiating between courses or up and down. The number of paces are then added together and an average is obtained. This method should give an average pace or number of paces per chain or tally which would be adequate for pacing without regard for slope. It would not give a true pace for level ground however.

Course	Up	Down
1	68	66
2	70	71
3	69	69
4	72	70
5	69	68
	Sum = 692	

Average number of paces per tally = 69.5
Average number paces per chain = 13.9 or 14.0
Number paces per 100 feet = 20.7 or 21

The pacer must be able to estimate distance by eye where the ground changes from the horizontal and have a good mental picture of the length of his horizontal pace so that it may be applied in steep and brushy terrain. Usually it is wise for the beginner to fill out the table of paces for practice, and as he becomes more proficient he will use the pacing table on long uniform, gradually-changing slopes and apply the ocular estimate for brushy, abruptly changing slopes.

This training aid is by no means infallible, for one cannot become proficient in pacing without constant practice. Slopes do not fall at even percents for uniform 5-chain distances, but knowing the number of extra paces required on these slopes, one has a better concept of the effect of slope on his pace, and he is able to adjust accordingly.

The pacing table should be copied into the field notebook where it will serve as a useful reference whenever pacing is required. Because the pace is merely another tool for measuring distance, the field notes for pacing will follow the same form as field notes for taping.

A constant rate of travel at a free and natural walking gait is the key to successful pacing. Consistency is developed only through continued practice, together with alertness on the part of the individual.

As previously stated, accuracy depends primarily on consistency. When one has become proficient at pacing, he should have an accuracy ratio of 1 in 80 or one chain in a mile in rough country. To retain this amount of accuracy will require constant practice and checking. For work requiring no more accuracy than 1 in 80, pacing is an inexpensive and rapid method of measuring distance.

Errors usually develop through one or more of the following:

a. A tendency to stretch out in the morning at a rate one is unable to maintain throughout the day.
b. Inaccurate recording.
c. Varied physical condition of the pacer.
d. Amount of brush.
e. Change in type of shoe.
f. Condition of soil and weather.
g. Walking with someone else. No two people will have exactly the same pace, and one is inclined to fall into step with the other person.

Since the natural tendency is to lose rather than gain accuracy, one should take advantage of every opportunity to check his pace over measured distances.

Two useful tools will improve the pacer's work. The first is the "tallywhacker," a counting device carried in the hand to help keep track of the distance. Pacers use this device to record distance by tallies, chains, or 100 foot lengths, and it may also be used to count each pace between recordings to the field notebook. The other aid is a stick or compass staff on which is marked the length of one pace. This is particularly handy when going up a steep hill or through heavy brush, when it is desired to visualize the actual pace.

Section 3. Problems

1. A line measured with a 100-foot tape is found to be 1246.80 feet long. Later the tape is found to be 0.04 feet too long. What is the correct length of the line?

2. With a 100 foot tape which is 0.07 feet too long, you are required to lay out a base line which is 1000.00 feet long. What distance would you measure to obtain the required distance?

3. You wish to lay off a line 8 chains 67.5 links long with a 200.0 foot engineer's tape which is 0.12 feet too short. What distance would you measure?

4. With a subtracting tape a distance was measured as being 82.30 feet. What number (a) was the rear chainman holding? (b) the head chainman?

5. A line was measured with a 100.0 tape and found to be 1005.46 feet in length Later the line was measured with a tape calibrated to be 100.03 feet long and found to be 1005.4 feet in length. Was the first tape used (a) long or short? (b) how much?

6. In measuring a line with a 100-foot tape, the zero end is held 3 feet too high. What is the error in one tape length?

7. A distance of 200 feet is measured along a 12 percent slope. What is (a) the horizontal distance? (b) the difference in elevation between the two points?

8. The difference in elevation is 76.4 feet; between two points along a 43 percent slope. What is (a) the horizontal distance? (b) the slope distance? (c) the vertical angle?

9. The distance along the slope is 135.0 feet. The horizontal distance is 127.5 feet. What is (a) the percent of slope? (b) the difference in elevation?

10. The horizontal distance is 283.6 feet and the difference in elevation is 38.4 feet. What is (a) the percent of slope? (b) vertical angle? (c) the slope distance?

11. The zero end of the tape was held 2.5 feet off line when measuring a distance of 180.0 feet with a 200.0 foot tape. What is the error in this measurement?

12. What slope distance would be measured with a 100-foot tape which is 0.06 feet too short on a 32.0 percent slope to lay off a line whose true horizontal length is to be 937.5 feet?

13. You wish to lay off a line which is 1 tally, 2 chains, and 7 links in length with a 200.0-foot tape which is 0.04 feet too short. What distance should you measure?

14. How many acres are there in a tract of land which is 12 chains on each side?

15. Between stations 17+50 and 30+30 a road has a uniform grade of 5 degrees. What is the (a) percent grade on this portion of the road? (b) what is the difference in elevation between these two points?

16. What would the head chainman and rear chainman hold respectively on a 200.0 foot subtracting tape is they were required to measure a horizontal distance of 171.42 on a slope of 63 percent if the tape were 0.3 feet too short?

17. The elevations of stations 3.45 and 6.80 are 512.5 feet and 597.0 feet respectively. What is the slope in percent between these stations and what is the slope distance?

18. A slope distance of 250.0 feet is measured between two points along an 18 percent slope. What is the horizontal distance and difference in elevation between these two points?

19. A line whose true length is 965.3 feet is laid off by measuring 966.1 feet with a 100.0-foot tape. Is the true length of the tape used more or less than 100.0 feet and what is the error?

20. If a 200.0-foot tape were permitted to sag 3.5 feet in the center while chaining horizontally, what would be the approximate error in this measurement?

21. The true length of a tape is 100.08 feet. What is the true length of a line measured to be 1096.52 feet? What distance should be measured with the same tape is the true length of the line is to be 894.70 feet?

22. The distance between two points was found to be 18 tallies plus 4 chain and 27 links. What is this distance in feet?

23. The difference in elevation between two points is 16.5 feet and the abney reading along this slope is -27%. What is the horizontal distance between these points?

24. You wish to lay off a line 1 tally, 2 chains and 56.5 links long with a 200.00 foot engineer's tape which is 0.11 feet too long. What distance would you measure?

25. What distance can a two chain tape be deflected out of line at the 1 chain mark on a slope of 0% before a distance error of 2 links will be observed?

26. Find the percent of slope when the horizontal distance equals 175.00 feet and the difference in elevation equals 27.6 feet.

27. A slope distance of 100 feet is chained on a 17% slope. What is the horizontal distance for this measurement? What slope distance must be measured on this slope to lay off 100 feet horizontal distance. (Hint: use table 1.)

28. What horizontal distance should be measured in addition to a slope distance of 100.00 feet on a 27 degree slope in order to obtain a horizontal distance of 100.00 feet? (table 2)

29. Find the true length of a 100 foot tape which gave a measured distance of 1905.62 on a line whose true length is 1906.50.

CHAPTER IV. MEASUREMENT OF DIRECTION

Section 1. Angular Relationships

Direction is the angular relationship between two lines passing through two points and intersecting or converging at a third point. This angle is usually made with a reference line called a meridian. The angle can be measured in degrees, minutes, and seconds. In compass work, the compass circle is divided into degrees, and the minutes in units of 15 must be estimated. Only with a transit or theodolite can seconds be measured.

Certain angular relationships must be understood when determining direction. These are:

1. True bearings are acute angles which a line makes with the meridian. Bearings are measured from the north-south line and can never exceed 90°. They are measured to the east or west of the meridian and are expressed as N ^{o}E, S ^{o}W, etc. Figure 12 illustrates the direction of lines in terms of bearings:

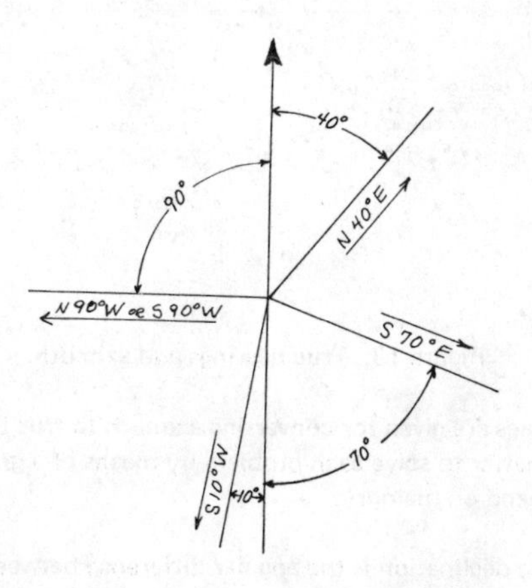

Figure 12. True bearings.

2. Azimuth is the angle that a line makes with the true north-south line and is measured from 0° to 360° in a clockwise direction. For almost all work done with a compass the azimuth is measured from North. However, for work with the Oregon State Coordinate system, azimuth is measured from South in a clockwise direction.

Northeast bearings are azimuths of 0 to 90 degrees; southeast bearings are azimuths of 90 to 180 degrees; southwest bearings are azimuths of 180 to 270 degrees, and northwest bearings are azimuths of 270 to 360 degrees. Figure 13 shows the relationship between true bearings and azimuths.

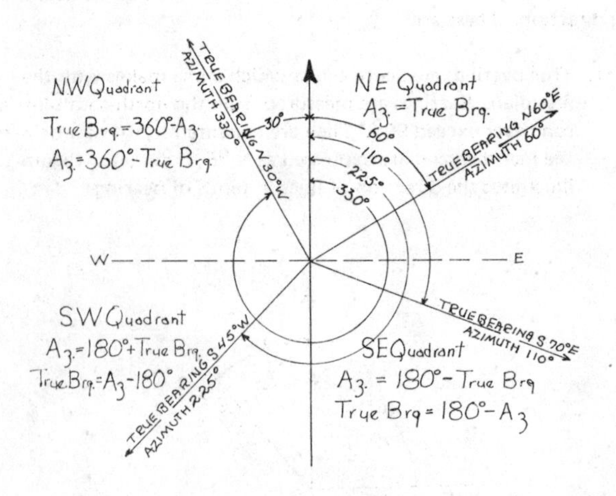

Figure 13. True bearings and azimuth.

While rules are given for converting azimuth to true bearings, etc., it is much better to solve each problem by means of a graphical sketch than to depend on memory.

Magnetic declination is the angular difference between true north and magnetic north. The earth acts like a great magnet. The source of magnetism is not concentrated at specific points but is distributed throughout the earth. The magnetic field in which the compass needle aligns itself is the surface evidence of all of the combined actions of all parts of the magnetic interior of the earth. Since the compass needle is magnetized, its direction is controlled by the magnetic lines of force at the surface of the earth. Hence magnetic north is the direction indicated by the north-seeking pole of the horizontal magnetic needle.

The magnetic poles are not important, but the North Magnetic Pole position was assumed to be 75 degrees north latitude and 101 degrees west longitude in 1960, and the position of the South Magnetic Pole as 67 degrees south latitude and 148 degrees east longitude. The poles are not points but large areas within which the direction of a compass can not be determined and within which a dip needle will stand vertical.

At the present time, in the eastern part of the United States, magnetic north lies to the west of true north, and this angle is referred to as west declination. In the western two thirds of the United States the compass, when pointing to magnetic north, will point to the east of true north, and, therefore, east declination exists in this section. The declination value may be found on an Isogonic Chart published by the U.S. Coast & Geologic Survey. Figure 14 is an Isogonic chart which is a map upon which lines are drawn connecting those points having the same declination, and these lines are called Isogonic Lines. The line connecting points having zero declination (area where the magnetic needle points true north) is called the Agonic line.

Isogonic Chart

This chart also shows rate of change in magnetic declination from year to year. In addition to a change of several degrees over a long period of time, the declination makes annual changes and even daily changes, but the amount is so small, it can be ignored in compass work. The declination of an area can also be determined from the true north-south line established by an observation on the North Star. However, local attraction might cause an error.

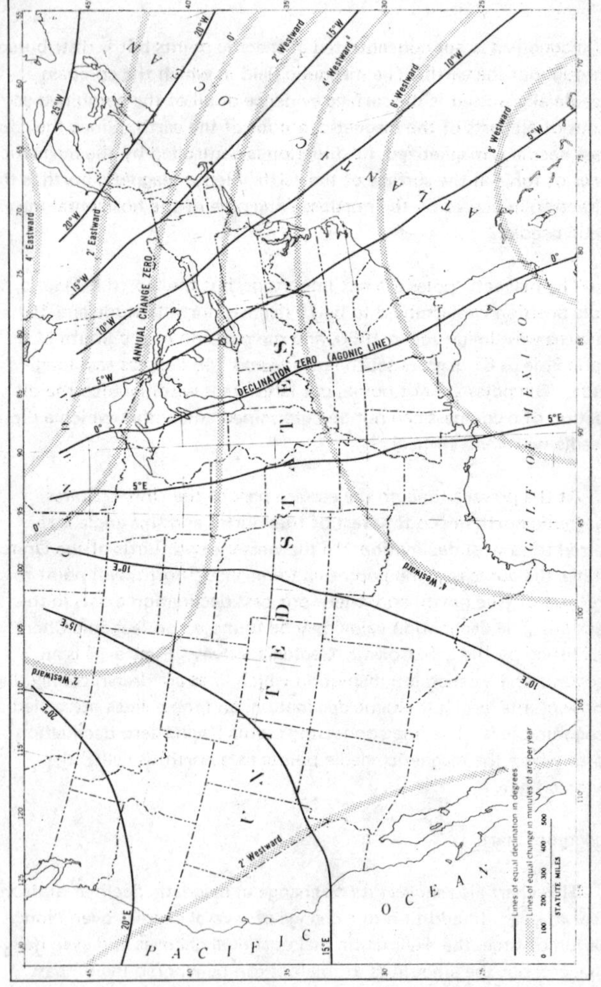

Figure 14. Isogonic chart of the United States, showing lines of equal magnetic declination. (Courtesy of Environmental Sciences Services Administration)

Magnetic bearing is the acute angle which a line makes with a magnetic north-south line. Whether the bearing read from the compass is a true bearing or a magnetic bearing is dependent on the declination being set off on the compass. If the magnetic declination is set on the compass (the compass scale rotated right or left for east or west declination respectively an angular distance equal to the declination) all bearings read will be true bearings. If the declination is not set off on the compass, the bearings observed will be magnetic bearings (Figure 15).

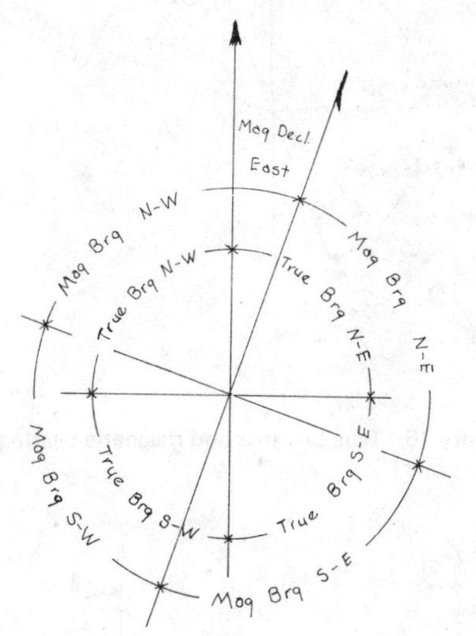

Figure 15. Relationship between true bearing and magnetic bearings.

Occasionally, it is necessary to convert true bearings to the magnetic bearing and vice versa. This is easily done if a sketch is used to clarify the situation (Figure 16).

By using the same approach, magnetic bearings may be converted to true bearings. This problem is often encountered in retracing old lines which have magnetic bearings and in areas in which the declination has changed since the running of the original line. To illustrate;

the observed magnetic bearing of a line survey in 1850 was N60°E, and the declination in 1850 was 12°E. What is the magnetic bearing of this line in 1967 if the declination in 1967 is 19°E? (Figure 17).

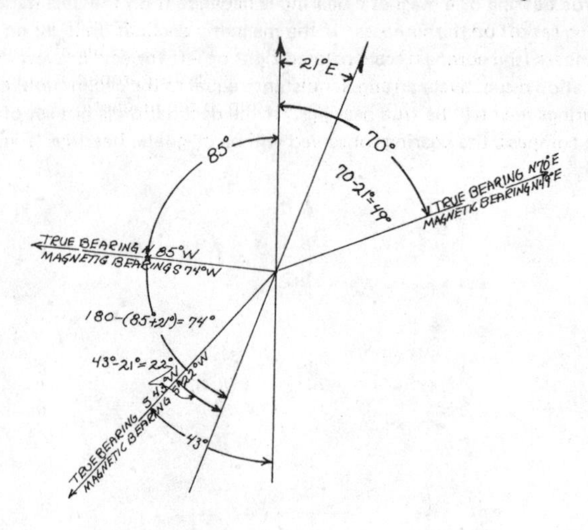

Figure 16. True bearings and magnetic bearings.

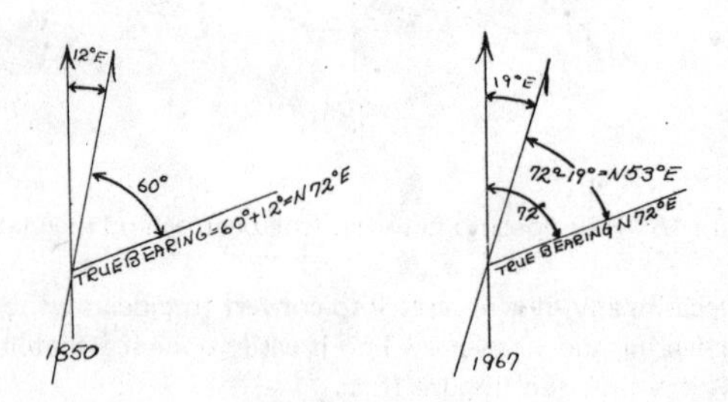

Figure 17. Corrections for change in magnetic declination.

Local attraction is a most common phenomenon occurring in mountainous areas particularly those of volcanic origin. The lines of magnetic force which orient the magnetic needle of the compass are frequently altered by ore and mineral deposits, fence lines, and electric transmission lines. Even such objects as choker cable and metal mechnical pencils in a shirt pocket will cause the compass needle to deviate from its normal oriented position in the magnetic field.

If local attraction does exist in an area in which a traverse is being run with a compass, it should be recognized that a local attraction existing will affect all compass bearings taken at any one point by the same amount.

Since bearings are a means of determining the values of angles, the values of the angles will not be affected if all of the bearings used to compute an angle are taken at the same point. Deflection angles and interior angles are the most common angular measurements in Forest Engineering. A deflection angle is the angle turned from the prolonged line through the station, or it is the angle between the prolongation of the backsight and the foresight, and it is always noted as being turned right or left from the prolonged backsight. An interior angle is the interior or inside angle between two adjacent sides of a closed traverse. The sum of the interior angles of a closed traverse is (n-2)(180) where n equals the number of sides in the polygon. The sum of the deflection angles in a closed traverse is 360 degrees when the right and left deflection angle are considered to have opposite signs. Figure 18 shows the relationship between interior and deflection angles.

In working with compasses a foresight is considered to be a sight taken on the following points or stations which may be indicated by a letter or number. A backsight is a sight or bearing taken on the preceding station. A foresight might be written as a bearing from C to D or from station 9+00 to 9+50. A backsight would be a bearing from D to C or from station 9+50 to 9+00. If there is no local attraction existing then, obviously, the foresight and backsight would be identical numerically; only the letters indicating quadrants would be changed.

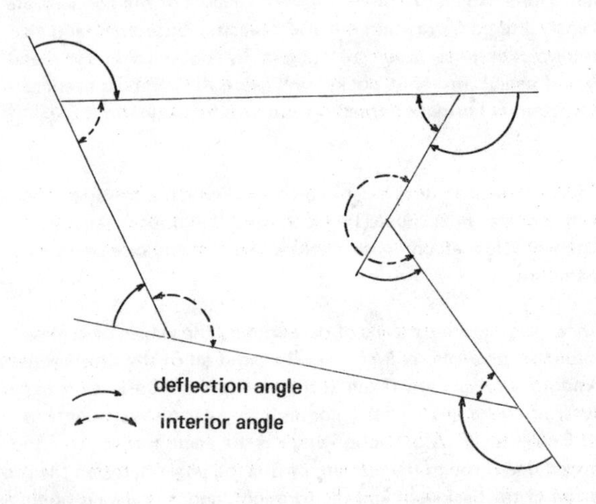

Figure 18. Traverse with deflection and interior angles.

If local attraction does exist, it is absolutely necessary to take foresights and backsights at every point on the traverse in order to be able to correct for local attraction and to calculate the correct bearing. Correcting or adjusting for local attraction is very simple providing that the backsight and foresight between two adjacent stations are the same. If no backsight and foresight agree, then it would be necessary to determine the bearing of one course by means of a celestial observation.

Figure 19 illustrates a method of adjusting for local attraction.

The arrow indicates the backsight and foresight bearings. The bearings AB and BA agree; therefore it is assumed that no local attraction exists at either station, and all bearings taken from B are unaffected by local attraction; hence, the bearing BC, which is N30^{o}00′E, is correct.

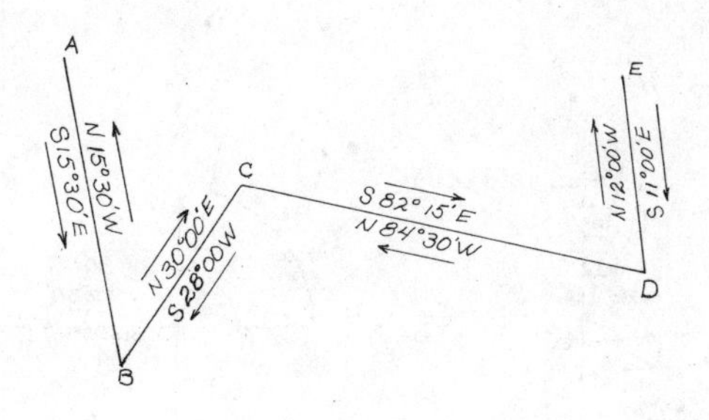

Figure 19. Local attraction in a traverse.

Find angle C or BCD.

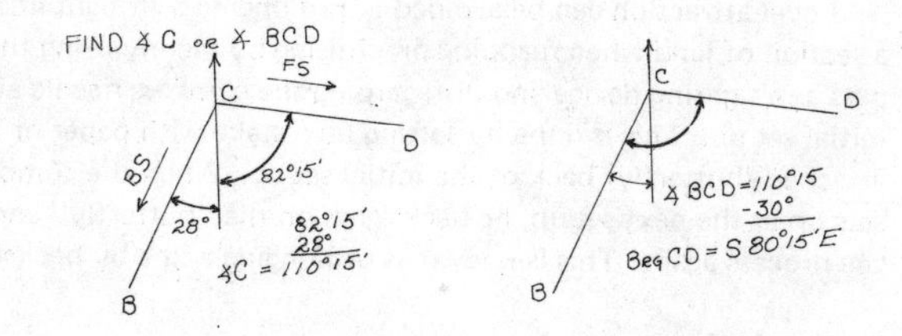

Figure 20. Correcting for local attraction at Station C.

If there were no local attraction at C, the backsight CB would have been S30°W instead of S28°W and the bearing of CD would be S80°15'E.

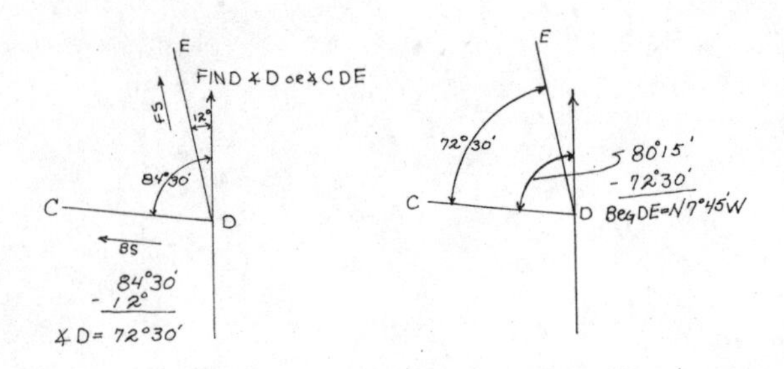

Figure 21. Correction for local attraction at Station D.

Assuming again that there was no local attraction at D, the backsight DC would then have been N80°15'W (bearing CD adjusted for local attraction equals S80°15'E), and the bearing DE would be N7°45'W.

Local attraction can be avoided in prolonging a straight line across a section of land when mapping or cruising by merely using the compass as a sighting device and disregarding the compass needle after the initial set up. This is done by setting up a stake with paper or flagging attached (butterfly) back of the initial set up. When the compassman sets up at the next set up, he backsights on the "butterfly" and repeats the process again. This is known as prolonging a line by backsights.

Section 2. Compasses

The essential features of a compass used in surveying are: (1) a circle graduated in angular units usually one degree dimensions, (2) magnetic needle, and (3) sighting device. Figures 22 and 23 illustrate two types of hand compasses. The box compass in Figure 22 has an etched sighting line through the north-south axis. The needle lifter is a plunger which raises the needle off its pivot when the lid is closed. The lifter is also convenient to use to stop the swing of the needle by tapping it with the finger, thus, settling it more quickly. The needle itself is of magnetized tempered steel, and supported by a jewel on the pivot. The north end of the needle is usually marked by an arrow, and the south end of the needle is identified by a small adjustable fine wire counterweight which can be moved to counteract the vertical dip of the needle.

Figure 22. Box or hand compass. (Courtesy of Leupold and Stevens)

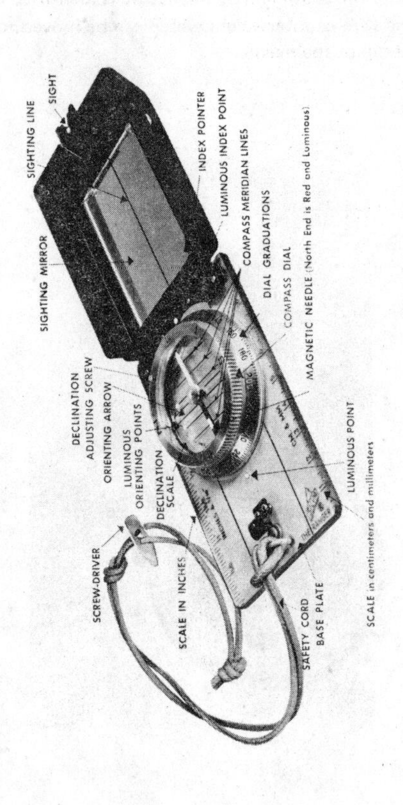

Figure 23. Silva hand compass. (Courtesy of Silva Inc.)

The Silva compass does not have a lifter because the needle is encased in a liquid which dampens both the vertical and horizontal movement of the needle. The dial of both types of compasses is adjustable so that the magnetic declination can be set off, making it possible to read true bearings instead of magnetic bearings.

The method of using the hand compasses is simple, and with little practice will provide a rapid method of traversing for less precise surveying work.

The hand compass is held with both hands about chest high with both arms against the body. To find a bearing of a given line, sight along the etch line of the box compass, and observe the bearing at the north end of the compass needle. To travel in a given direction, rotate the whole body holding the compass in front of the body until the north end of the needle points to the direction in which it is desired to travel. Sight along the etch line on the lid of the compass, pick out an object on the line ahead, and "follow your nose."

To use the Silva compass, the lid, which contains a mirror for viewing the compass needle, is tilted or raised up to find the bearing of a line; then the compass is held in the right hand at almost eye-height and a sight taken on the line. The left hand then rotates the compass dial until the compass needle is centered in the orienting arrow etched on the glass. The bearing of the line is then determined at the index pointer. To travel in a given direction using a Silva compass, one first rotates the compass dial and the desired bearing is then set on the index pointer. Holding the compass squarely in front of the body, the body is rotated until the magnetic needle is centered in the orienting arrow. A sight is then made over the front U sight, and an object is picked which is on line. While doing this sighting, it is also necessary to observe the compass needle in the mirror to make certain of its position with respect to the orienting arrow.

Staff Compass

The staff compass or surveyor's compass consists of a box compass similar to the hand compass. The sight is accomplished by sighting through two vertical sighting vanes set directly opposite each other on the north-south axis. The rear sight consists of a narrow aperature. The front sight contains a vertical hair or metal band mounted at the center of a wide opening. The compass is attached to a pipe-like fitting which slips onto a spindle which, in turn, is attached to a ball in a ball-and-socket joint. The upper part of the socket is a thumb-nut which is tightened until the ball is securely held by friction. All of these fittings known as the leveling head are attached to a wooden staff, sometimes called a Jacob's Staff, Figure 24.

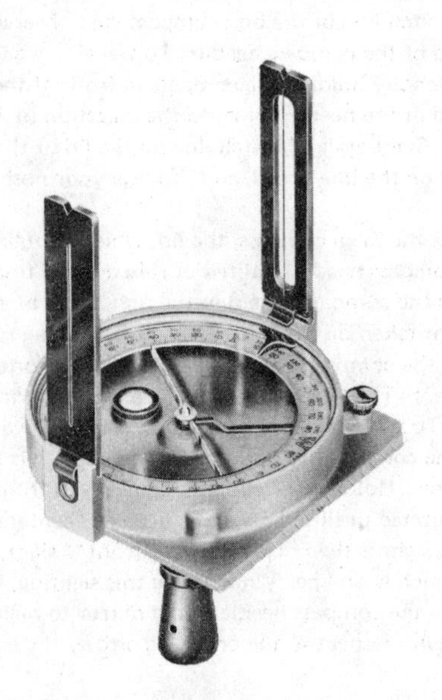

Figure 24. Staff compass. (Courtesy of Warren Knight)

After the staff is placed firmly in the ground, the compass is attached to the leveling head and leveled by means of either a circular level vial (bull's eye or fish eye) or two level vials mounted at right angles to each other. The staff compass also has a thumb screw for lifting the compass needle off its pivot. The declination is also changed by means of an exterior screw.

Declination adjustments are often needed in the field to correct errors or to compensate for changes in declination. An angle of one degree subtends a distance at one mile of approximately 92 feet. That is to say $x = 92$ feet when $\frac{x}{5280} = \sin 1^{\circ}$ or $\tan 1^{\circ}$.

Knowing this relationship, the declination can be corrected accordingly by prolonging a straight line from one corner to another and then determining the amount of line error. The error divided by 92 gives the changes in degrees needed to correct the declination. After changing the declination, the line is rerun in the opposite direction to check the correction.

Care must be taken to make the declination correction in the correct direction. In areas in the United States where the declination is East, increasing the declination will throw the line to the left while decreasing the declination will move the line to the right. For example, assume that in Oregon the declination is 20°E. With no declination set off on a compass, a bearing of N 0°E would take you toward magnetic north but with a declination of 20°E set off on the compass, a bearing of N 0°E would take you in a true north direction. The direction of travel has shifted to the left as the declination has been increased from 0°E to 20°E.

Accuracy in the use of the hand compass in connection with pacing is about 1/80. With the staff compass, which is usually used in connection with tapes, the accuracy ratio is about 1/300.

If the compass is not in proper adjustment, it is best to return it to the company rather than attempt to repair or adjust it yourself. If the equipment is properly cared for, it will give many years of satisfactory service.

Section 3. Problems

1. The magnetic bearing of line AB = N 28° 30′ E. The true bearing of AB is N 14° 45′ E. What is the magnetic declination?

2. The bearing of line AB is N 20° 15′ E. The bearing of line BC is S 61° 30′ E. What is the deflection angle at B in running a traverse from A to B to C?

3. Given: Traverse ABCD
   Magnetic declination = 19° 30′E
   All bearings given are true bearings

   Find Corrected:
   - Bearing AB
   - Bearing BC
   - Bearing CD
   - Azimuth CB
   - Deflection angle at C
   - Azimuth DC

4. Using a compass on which the magnetic declination of 20° 30′ E cannot be set off, you wish to lay out a regular pentagon in a clockwise direction in such a manner that the true bearing of the first course is N 86° 30′ W. What magnetic bearings would you use?

5. You run a line one mile in length N 90° E and end up 69 feet north of a point known to be due east of your starting point. If the declination set off on your compass was 21° 30′ E, what declination should you have used? Assuming that the declination could not be changed, what bearing would you use in retracing the line from point to point in the opposite direction?

6. Compute the missing data.

True Bearing	Declination	Magnetic Bearing	Azimuth
N 29° 30′ W	———	N 45° 15′ W	——————
S 86° 15′ E	———	N 84° 00′ E	——————
S 64° 00′ W	———	S 41° 30′ W	——————
N 01° 30′ E	———	N 19° 45′ W	——————
————————	19° E	N 52° 30′ E	——————
————————	20° E	N 81° 15′ W	——————
————————	21° W	S 76° 00′ W	——————
————————	18° W	S 04° 45′ E	——————
————————	6° W	————————	347° 15′
————————	12° E	————————	265° 30′
————————	20° W	————————	162° 45′
————————	14° E	————————	26° 30′

7. The following deflection angles were turned at stations 1+00, 2+00, 3+00, 4+00, and 5+00; 16°L, 11° 30′ R, 8° 15′ R, 4° 30′ L, and 82° 45′ R. If the calculated bearing from 0+00 to 1+00 was N 12° 15′ E, compute the bearing between each of the remaining stations.

8. The interior angles of a closed traverse were as follows: A – 116° 15′, B – 89° 45′, C – 103° 30′, D – 128° 45′, and E – 101° 45′. If the bearing from A to B is N 87° 30′ W, compute the bearing of the other courses.

9. Bearing AB = N 68° E
Bearing AC = S 43° E
Distance AB = 150.0 feet
Distance AC = 220.0 feet
Find: Bearing BC, Distance BC and area ABC.

10.

Course	Foresight	Backsight
AB	S 72° 15′ W	N 71° 00′ E
BC	S 10° 30′ W	N 13° 45′ E
CD	S 68° 00′ E	N 66° 30′ W
DE	N 49° 45′ E	S 51° 15′ W
EF	N 30° 30′ W	S 30° 30′ E

Compute the corrected bearing of line AB.

11. The bearing from D to E = S 63° 30′ E. At E a left deflection angle of 38° 15′ is turned. What is the bearing of EF?

12. The mangetic bearing of a line was found to be N 42° 45′ W when the declination was 6° 30′ W. What magnetic bearing would be used to retrace the line if the declination has changed to 2° 45′ E?

13. The true azimuth of a line is 14° 15′ while the magnetic azimuth of this line is 352° 30′. What is the magnetic declination?

14. If the declination set off on your compass was 21° 30′ E instead of the correct value of 20° 00′ E, how much line error would you expect in a distance of a ½ mile? Would the line error be right or left?

15. Magnetic bearing CD = _____ Bearing BA = _____
Corrected azimuth CB = ____ Magnetic bearing DE = _____

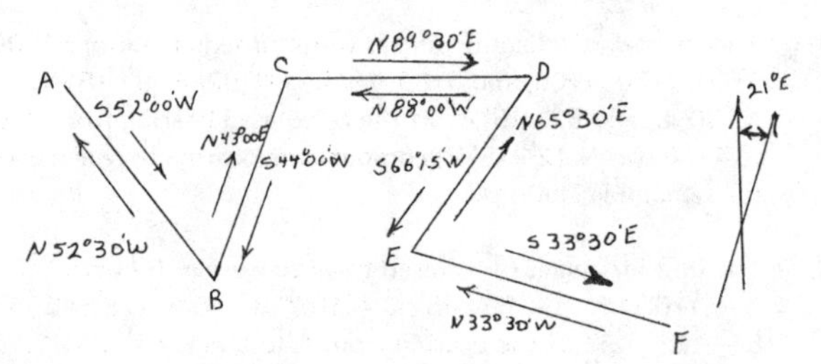

16. The magnetic bearing of a line was N43°30′E when the declination was 15°00′W. Find the magnetic bearing of the same line if the declination has changed to 11°30′W.

17. If the magnetic declination is 20°15′E, find the magnetic bearing of line which has a true bearing of S16°45′W.

18. The azimuth of a line is equal to 237°30′. If the magnetic bearing of this line is S39°45′W, what is the magnetic declination?

19. Each deflection angles of a regular polygon run in a clockwise direction is equal to 30 degrees right. Find the sum of the interior angles.

20. The magnetic bearing of a line in 1864 was determined to be N46°30′W when the declination was 3°30′E. In 1973, the magnetic bearing of the same line was found to be N37°45′W. What was the magnetic declination in 1973?

CHAPTER V. MEASUREMENT OF VERTICAL DISTANCES

Section 1. Aneroid Barometer

The barometer is an instrument for measuring the pressure of the atmosphere. A reliable method of measuring this pressure is the mecurial barometer which measures the pressure in inches of mercury. This equipment is not portable; hence, the aneroid barometer which is portable is often used for less precise measurements in forest surveying. It consists of a vacuum box with one very flexible sensitive side, which moves in or out as the air pressure changes. This flexible movement is attached by means of a system of linkage and levers to a pointer on the instrument's dial. The dial is calibrated to read the pressure in the number of inches of mercury raised by this pressure and also the elevation in feet corresponding to this pressure under normal conditions. Figure 25 shows a pocket size barometer or altimeter, as they are sometimes called. Since the air pressure decreases as the elevation above sea level increases, the change in pressure corresponds to a change in elevation. However, the air pressure will vary and change continuously at the same elevation. Hence, the elevation observed with an aneroid barometer is not always accurate. This instrument is most useful in determining the difference in elevation between points over a short period of time. The aneroid barometer does have a place in forest surveying and is used in reconnaissance work with the hand compass in conjuction with pacing to measure roughly the difference in elevation between two control points which will control the percent of slope of a proposed logging road. The best results are obtained when the barometer is checked or tied into points of known elevation every two hours. The aneroid barometer is a delicate instrument and should be protected from sharp blows and extreme heat. When using it in the field, the following precautions should be observed:

1. Tap the instrument gently to check on the free movement of the working parts.
2. Wait a few minutes before reading at a new point to allow the lag in needle to adjust to the change in pressure.
3. Always read in the same position.

Figure 25. Pocket size aneroid barometer. (German made)

As previously mentioned, the readings are affected by the changing air pressure. There are several methods of making corrections to the observed readings, but only one will be described in detail. This method requires that two readings be taken at the initial point of known elevation. This known elevation is set on the barometer by adjusting the dial so that the pointer indicates the proper elevation in feet. The time of this reading is noted. Elevation of successive points are recorded, and the time of each observation noted to the nearest five minutes. The last observed elevation is taken at the initial point, and the time is again recorded. If there is a difference between the first and last readings, which would indicate a change in air pressure, and the elapsed time is not too great, it can be assumed that the change in pressure is proportional to the elapsed time. The elevation corrections to be applied to any observed elevation is, therefore, equal to the difference in elevation between the first and last reading multiplied by the elapsed time to reach that point divided by the total elapsed time between the first and last reading. In other words, the correction is equal to the error, times the elapsed time, divided by the total elapsed time. The proper sign must be attached to the correction before applying. If the error is plus, which would indicate that all the observed elevations are too high, all of the corrections then would be subtracted from the recorded or observed elevation. If the error is minus, then a plus sign would be in order for all of the corrections. In recording notes in the field book, the notes are started at the top of the page and read down. Only in leveling are the notes recorded in this manner. In all other situations, the notes begin at the bottom of the page and read up. Figure 26 illustrates the note form for aneroid barometer field notes.

	Aneroid Barometer			
Sta.	Time	Obs. Elev	Elev Correction	Corrected Elev.
Foe. Bldg	8:00AM	350′		350′
Oak Creek Saddle	8:30AM	800′	−40	760′
Center Sec. 8	8:45AM	650′	−60	590′
E¼ Cor. Sec. 8	9:30AM	1100′	−120	980′
Foe Bldg	10:00AM	510′	−160	350′

Figure 26. Aneroid barometer note form.

Section 2. Clinometer

The instruments used for measuring slopes or for indirect leveling are the abney level or clinometer. The latter does not need adjusting and is very easy to use. It is held to the eye and moved in a vertical arc until the sighting line, viewed through the lens is aligned on the desired object. The angle is read by looking through the lens while sighting on the object alongside the clinometer. Figure 27 is a Suunto Clinometer and Figure 28 shows the clinometer as it would appear held up to the eye. The scale which is on a wheel shaped assembly is supported by a jewel bearing and immersed in a liquid to dampen its movement.

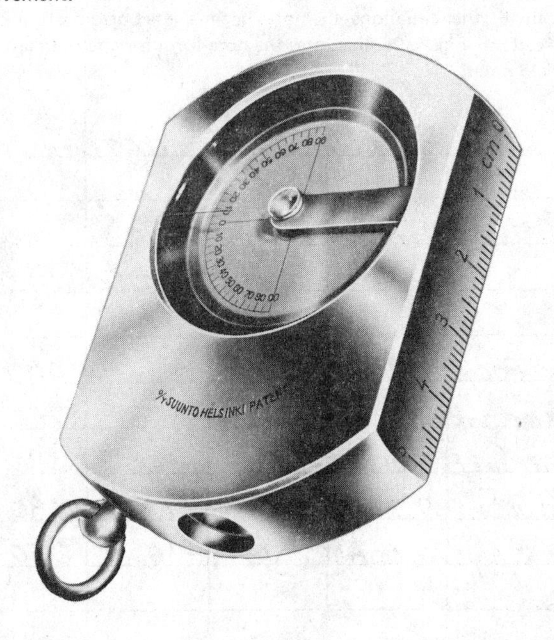

Figure 27. Clinometer. (Courtesy of Suunto)

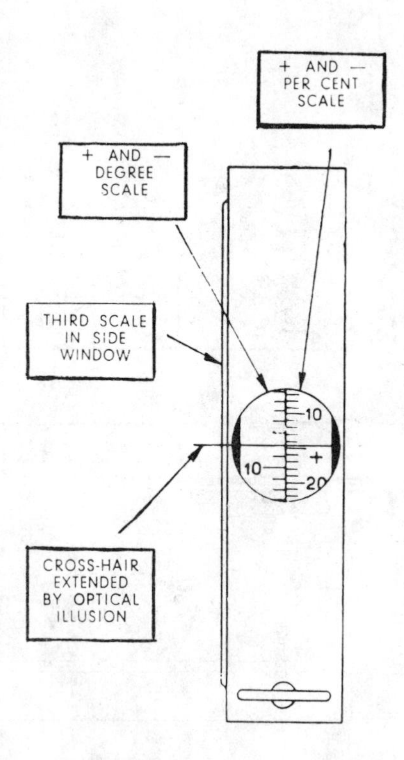

Figure 28. Eye view of clinometer.

Section 3. Abneys

The abney level is an instrument used for measuring slopes or for direct leveling (see Figure 29. The instrument consists of a square metal sliding tube on which a vertical arc is mounted with an index arm holding a level vial. Inside the square tube opposite the center of the arc is a mirror or prism which reflects the image of the bubble in the level vial to the eye. A cross-bar or cross-hair is mounted horizontally inside the sliding tube opposite the "I" end. The level vial

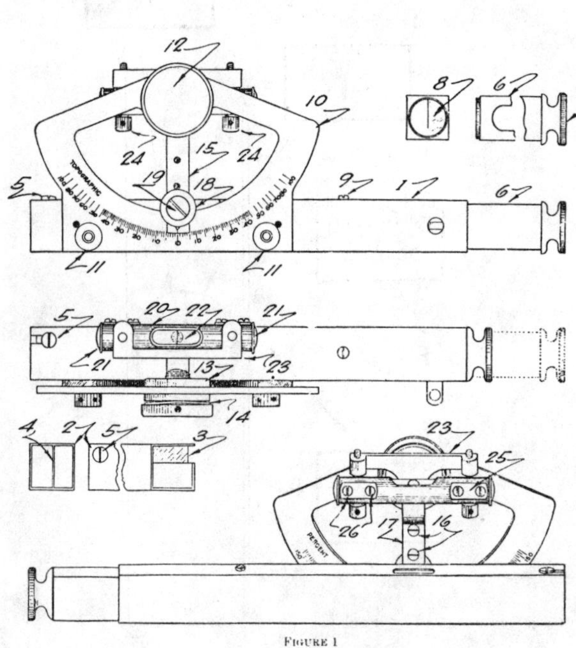

FIGURE 1

1. Telescope tube.
2. Prism and cross-hair slide.
3. Prism.
4. Cross-hair slide.
5. Prism slide lock screw.
6. Slide tube.
7. Eyepiece cap.
8. Half lens.
9. Slide tube lock screw.
10. Graduated limb.
11. Limb capstan nuts.
12. Main capstan bolt.
13. Main clamp nut.
14. Spring washer.
15. Index arm.
16. Index arm screws.
17. Index arm lock bar.
18. Index arm lock nut.
19. Lock nut screw.
20. Vial tube.
21. Vial tube ends.
22. Glass vial.
23. Vial bracket.
24. Capstan adjusting screws.
25. Capstan screw anchor.
26. Capstan screw anchor screws.

Figure 29. Abney hand level.

is attached to the index arm. It can be rotated about the horizontal axis of the vertical arc. The amount of rotation is then read on the arc at the mark on the index arm. Some abneys have an etched line on the level vial which should coincide with the horizontal bar at the end of the abney. Other abneys will have no etched line on the level vial. They will have only the horizontal bar. A third type of abney will have neither; it will have only an etched line on the prism which reflects the level vial and the level bubble to the eye. The adjustment of these various types of instruments will be discussed under the two-peg method of adjustment. The abney is an instrument for measuring the angle between the horizontal plane and the slope. The horizontal plane represented by the axis of the level vial, and the slope is represented by the line of sight which must be taken parallel to the slope. In other words, the observer rotates or moves the index arm until he has the level bubble in the vial centered on the cross-hair line. The value read on the arc will then give the slope. The slope which is the angle between the horizontal plane and the line of sight, may be read in one of three different units. Most abneys come complete with interchangeable arcs which are graduated in one of the three following manners:

1. Percent – one unit on the arc represents a change of 1 foot in elevation for each hundred feet of horizontal distance.
2. Topog – one unit on a topog represents a change of 1 foot in elevation for each chain or 66 feet in horizontal distance.
3. Degrees – one unit on this arc is equal to 1/360 angular part of a circle.

The most important adjustment necessary for an abney is the two-peg method which is a check to make certain that the level vial is parallel to the line of sight. Before making this adjustment, it is necessary to check and make certain that the horizontal bar in the slide at the end of the abney or level tube coincides with the etched line on the level vial, if the vial is so marked. If they do not coincide, they can be made to do so by adjusting the slide which contains the horizontal bar and the prism. This adjustment is made by moving in or out the slide which contains this horizontal bar. If there is no etched line on the level vial, this check is not important. The height of the level vial above the level should be checked so that on very steep slopes the etched line should appear near the concave side of the bubble, which is crescent shaped because of refraction. To adjust the abney by the two-peg method (Figure 30), select two objects such as two trees or two posts about 50 feet apart. Since there is no magnification, in the abney, it is not easy to sight on points at a greater distance. On the other hand, two points too close together will not insure as good an adjustment of the abney.

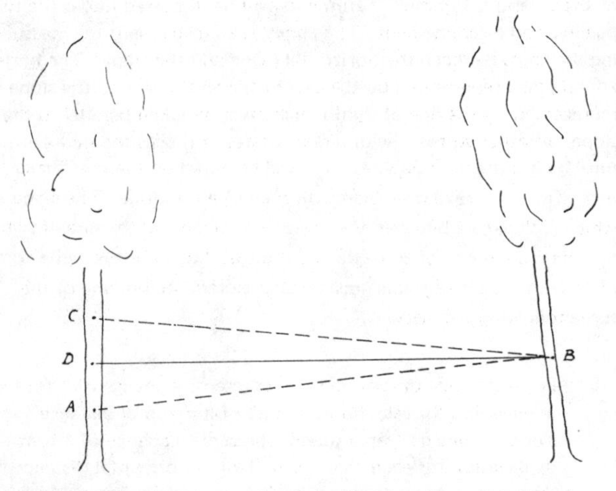

Figure 30. Two peg method of adjusting abney hand level.

First, set the abney on zero and sight from point A on one tree to the second tree and mark this point B. Move to point B and making sure that the abney is still set on zero, sight from point B back to first tree and mark this point C. If the abney is in good adjustment point C and A will coincide. If the abney is not in adjustment, then C and A will differ. Select a point D midway between A and C and with the abney held on point D, sight on point B. When this is done, the level bubble will no longer be centered on the cross-bar. To adjust the abney, raise or lower one end of the level vial by turning the level vial capstan nuts until the bubble again is centered while sighting from point D to point B. To check this adjustment, it is necessary to move to B and again sight on D to be sure that the adjustment made was the proper amount. The precision which can be obtained with the abney, of course, depends upon the experience of the crew using the equipment. Double abney work, which is the situation when the head and rear chainmen each use an abney, tends to eliminate some error and results in more accurate work. For ordinary single abney work, the accidental error that accumulates should not exceed 10 feet in elevation per mile. The accuracy of the abney is comparable to that of the staff compass. Hence, the two are used extensively in combination when mapping.

The source of error with the abney, of course, is due to the man using the instrument. The most common errors are: (a) abney improperly adjusted; (b) failing to hold the instrument steady when taking a reading; (c) calling a plus reading minus and a minus reading plus on a small slope; (d) sighting to a point which is not the same height above the ground as is the observer's eye. To eliminate sighting errors, a flashlight is almost a necessity to taking good sights in brushy areas.

Abneys with no etched line on the level vial can also be adjusted by the two-peg method by moving the slide in and out instead of adjusting the level vial itself. This is possible since the slide also contains the prism, and the movement of the slide will change the relative position of the sighting bar on the level vial.

There is almost no limit to the uses of the abney. Figure 31 illustrates how it is used to determine the heights of trees. The horizontal distance must be known and expressed in units of 100 feet. Since $\% = DE/HD \times 100$ and % is the abney reading it follows that the $DE = \frac{\%(\text{abney}) \times HD}{100}$. If a topographic abney is used, the base horizontal distance becomes one chain or 66 feet.

The abney can be set on zero and used as a hand level. It is used to run grade lines for roads by setting it on a predetermined reading. The grade line is established by moving the head chainman up or down slope until the desired reading is obtained. In this type of abney work the horizontal distance is not important because in measuring the slope, the angle between the horizontal and the line of sight is independent of everything else.

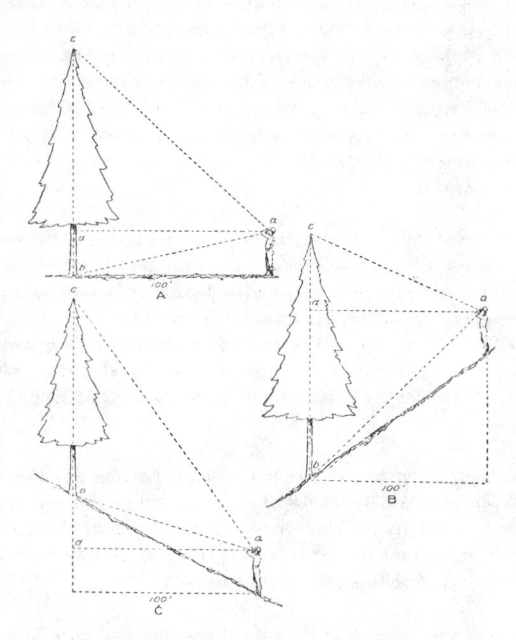

Figure 31. Measuring tree heights with percent abney.

Quite often distances are more conveniently measured on the slope than on the horizontal, particularly when the slope exceeds five percent. For any slope under five percent, for all practical purposes, the slope distance and the horizontal distance are considered to be equal to each other, but a difference in elevation still exists. When using a percent abney, the engineer's tape in 100-200 or 300 foot lengths is ordinarily used. Since the horizontal distance is less than the slope distance and the difference in elevation is determined by the slope and the horizontal distance, tables are necessary to convert the slope which is measured in percent and the slope distance to horizontal distance and difference in elevation. These tables are found in the Appendix and include slope distances in five-foot intervals. The slope tables are prepared for slopes varying from 6 percent to 80 percent. For any slopes exceeding 80 percent, the table then gives the value of this angle in terms of degrees and also the sine and the cosine function. The tables thus prepared merely convert the slope distance to horizontal distance by multiplying the slope distance by the cosine of the angle and to difference in elevation by multiplying the slope distance by the sine of the angle. Since the slope is read in percent, the tables are entered under the proper percent heading. The first column to the left under each percent heading gives the corresponding horizontal distance for the measured slope distance for that percent of slope. The second column under each percent heading gives the difference in elevation for that slope distance and percent of slope. While it is possible to interpolate directly in the tables for reading other than those given, a much simpler method is explained as follows.

Example: Use an abney reading of 62% for each example:

Slope distance (feet)	Look opposite	HD	DE
a. 27.5	27.5 = 1/10 x 275		
	∴ take 1/10 of HD & DE	23.37	14.49
b. 137	100 & record	84.99	52.70
	30 = 1/10 of 300		
	∴ record 1/10 of HD & DE	25.49	15.81
	7 = 1/10 of 70		
	∴ record 1/10 of HD & DE	5.95	3.68
		116.43	72.19
c. 321.4	300 and record	254.97	158.09
	20 = 1/10 of 200	17.00	10.54
	1 = 1/100 of 100	.85	.53
	(21 = 1/10 of 210)	(17.85)	(11.07)
	.4 = 1/100 of 40		
	∴ 1/100 of HD & DE	.34	.21
		273.16	169.37

The use of the topographic abney is exactly the same as the percent abney, the only difference being in the construction of the scale on the arc. The topographic abney is used with the topographic or surveyor's tape, and it measures the difference in elevation per chain or 66 feet horizontal distance. The topographic tape also contains a trailer which makes it possible to measure on the slope or ground slope a distance which is equal to one chain or two chains horizontal distance. The slope distance is the hypotenuse of a right triangle. As the angle of the slope increases, so does the abney reading and the slope distance. It is upon this principle that the trailer is graduated. With the two-chain tape, there are two trailers, one for use with the one chain which is marked on the underside of the tape. Figure 9c shows the graduation of the tape in links in solid numbers; while the trailer divisions are shown as broken numbers, as they appear on the other side of the tape. Figure 9d illustrates the trailer for the two-chain tape. The method of using the trailer is the same and the procedure followed is the same whether the abney reading is plus or minus.

To use the trailer, the head chainman goes up or down the slope a distance of one or two chains. Abney readings are then taken and if taken properly, should agree with only a difference in sign The chainmen having taken the abney reading, now chain a slope distance of one or two chains plus a distance on the trailer which corresponds to the number read on the abney.

The rear chainman merely lets the head chainman move ahead an amount equal to the number read on the abney. If 30 were read on the abney, the rear chainman would hold 30 on the tape beyond the 1 or 2 chain mark depending on whether a horizontal distance of 1 or 2 chains is desired (Figure 32). To eliminate any possible error, the abney readings should be taken again after the head chainman has marked his point to make certain that the abney reading still agrees with the amount of trailer used. It is not necessary to know the slope distance because the distance now marked will be one or two chains apart. In using the two-chain trailer, it is not necessary to double the amount of trailer or the number which was read on the abney, because this correction has been calculated for this trailer. The topographic abney and tape permit faster and cheaper work when it is possible to measure all the distances to one or two chain units. The work is slowed up and becomes more intensive when it is necessary to make frequent measurements and fractional distances or distances which

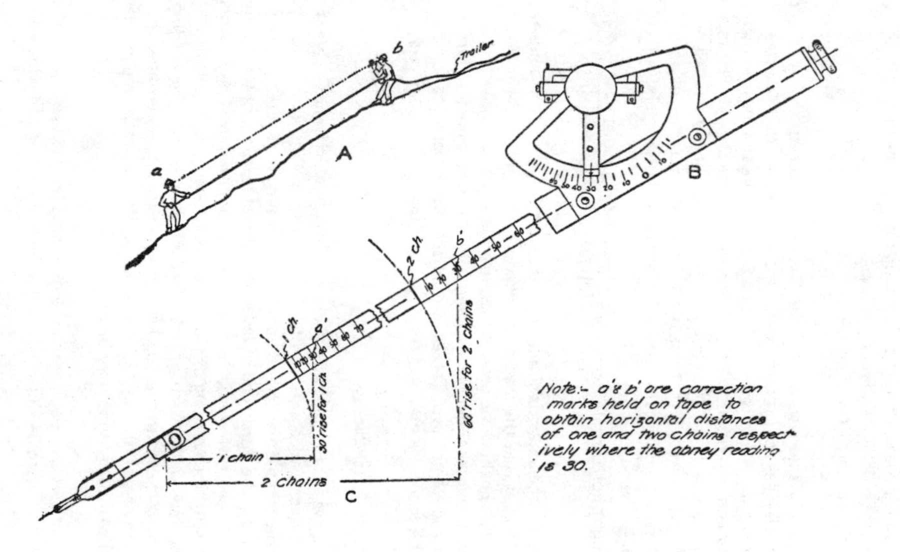

Figure 32. Slope chaining with topographic tape and topographic abney.

are less than one chain. Fractional slope distances are reduced to horizontal and differences in elevation by one of three methods:

1. The tables which are included in the back of the book.
2. Direct measurement or breaking tape and measuring horizontal distances directly.
3. By means of the formula.

The formula is derived from the theorem that corresponding parts of similar triangles are proportional. If it is necessary to measure a slope distance which is less than one chain, make the measurements directly and determine the slope by means of the abney. The tape is then turned over, and the actual distance to the abney reading on the trailer is observed. This, then, gives the slope distance that you would have measured had you gone one chain horizontal distance. Hence:

$$\text{Sought horizontal distance} = \frac{\text{Measured slope distance}}{\text{Slope distance for 1 chain horizontal distance}}$$

The degree abney is similar to the topographic and the percent abney except that the graduation marks on the arc are different. This abney enables the user to measure the slope in degrees. It is not as popular as the other two and is not used to a great extent in the Pacific Northwest. Tables in the back of the book are for use with the percent abney and the engineer's tape.

Section 4. Leveling

There are three methods of determining the difference in elevation between two points on the surface of the earth. Two of these methods, trigonometric and barometric leveling have already been discussed. The third type of leveling is direct leveling, of which there are two types, differential and profile. Differential leveling is concerned with the carrying of elevations from one point to another without regard for elevations of intervening points. Profile leveling, a type that is used in road design and construction, determines the elevation of stations at regular intervals so that, as the name implies, a profile of the ground may be plotted.

Direct leveling requires the use of a hand level (Figure 33) and a level rod. Figure 34 illustrated a type of self-reading rod which is

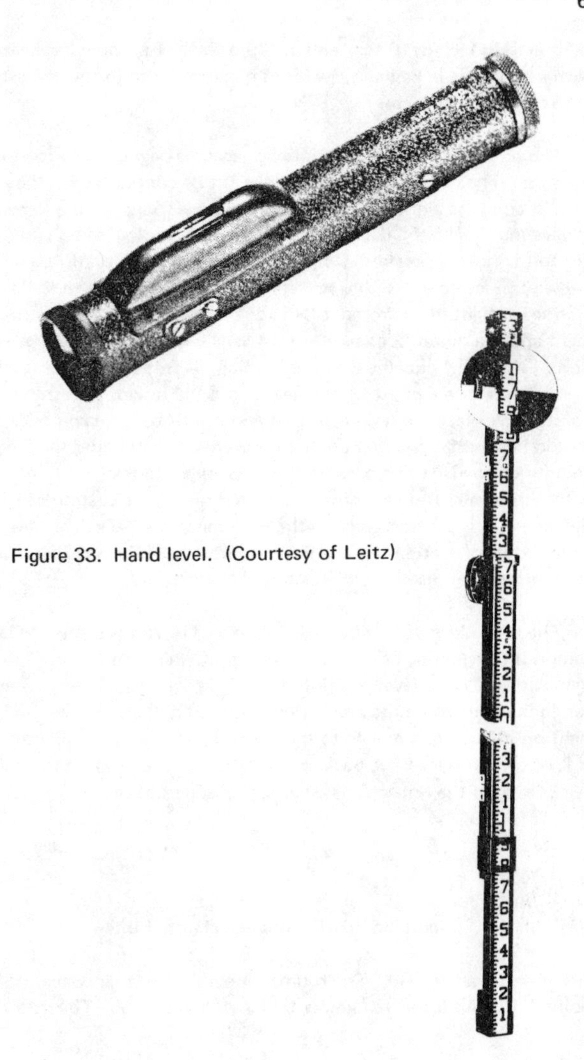

Figure 33. Hand level. (Courtesy of Leitz)

Figure 34. Level rod. (Courtesy of Keuffel & Esser Co.)

calibrated in feet, 1/10 ths., and 1/100 of feet. For most level work with a hand or Locke hand level, rod readings taken to the nearest 1/10 foot are satisfactory.

The note form shown in Figure 35, again is begun at the top of the page. The column headings should not be confused with those used in compass work even though the abbreviations are the same. In leveling, backsight (B.S.) is a sight or rod reading taken while the rod is held on a station or Bench Mark whose elevation is known. In leveling, B.S. has no reference or relationship to the di- :tion. Height of instrument (H.I.) is the elevation of the instrument above a datum such as mean sea level and is equal to the elevation of the point plus the B.S. rod reading. A foresight (F.S.) is a rod reading taken on a point whose elevation is unknown. A turning point is any place the level rod is set and rod reading taken on it. It should be emphasized that a value is never recorded under the F.S. column on the first line because it is impossible to take a F.S. on a point whose elevation is known. To determine the elevation of the T.P. the F.S. is subtracted from the H.I. Since the backsights are added to the elevation and the F.S. are subtracted from the H.I., the column headings used are (+S) and (–S) respectively

The procedure of leveling is as follows: The rodman sets the level rod on the beginning point, the elevation of which is known. The man with the hand level positions himself in such a manner that he can look at the rod in its present position but will also be able to sight on it when it is moved to the first T.P. without changing the H.I. of the hand level. A backsight is taken on the rod and this value is recorded in the notes. The backsight is added to the

Sta.	B.S.	H.I.	F.S.	Elev.
B.M.	7.5	848.0		840.5

Figure 35. Differential level note form.

elevation to get the H.I. The rodman moves ahead to a convenient point T.P., and a rod reading is taken with the level. This is a foresight on T.P.$_1$. The F.S. is subtracted from the H.I. to obtain the elevation of T.P.

Sta.	B.S.	H.I.	F.S.	Elev.
B.M.	7.5	848.0		840.5
T.P.$_1$			2.3	845.7

The rodman keeps the rod on the same point T.P., the man with the level moves ahead to a convenient point and again sights on the rod with the level. This is a B.S. and since it is on T.P. it is recorded in the proper column. This backsight is added to the elevation of T.P. to give an H.I.

Sta.	B.S.	H.I.	F.S.	Elev.
B.M.	7.5	848.0		840.5
T.P.$_1$	9.3	855.0	2.3	845.7

The rodman moves ahead again and another sight (F.S. on T.P.) is taken and recorded. The man with the level moves ahead with the level and sights back on the rod (B.S. on T.P.). This method of leap frogging is continued until the elevation of the last station is determined.

A quick check on accuracy of all work can be made. The difference between the sum of the foresights and the backsights must equal the difference in elevation between the beginning and the end point. Always check to see that it does. If it does not, recheck the addition and subtraction, and if an error still exists, it is best to re-run the level traverse again.

Section 5. Problems

1. What abneys readings would be obtained under the following conditions:

Horizontal distance	117.5 feet	160.0 feet	
Difference in elevation	28.4 feet		37.8 feet
Slope distance		182.0 feet	245.0 feet
Topographic abney	______	______	______
Percent abney	______	______	______
Degree abney	______	______	______

2. The following data was obtained with an aneroid barometer.

Station	Time	Observed Elevation
A	8:00 A.M.	1600 feet
B	9:30 A.M.	1750 feet
C	10:15 A.M.	1900 feet
D	10:45 A.M.	2100 feet
A	11:30 A.M.	1850 feet

Find the corrected elevation of station C in feet. Find the minimum length of road (horizontal distance) necessary to connect station A and D on a uniform 5½% grade.

3. The abney reading on a slope is +18½%. What slope distance would be required to reach a point 15 feet higher than the station from which the abney reading was obtained?

4. In tying into a corner, you measure the slope distance and get 86 links and with a topographic abney read +45. Find the horizontal distance and the difference in elevation.

5. The difference in elevation between two stations on a traverse is -37.5 feet and the abney reading with a percent abney on this slope is -29½%. What is the slope distance between these stations in chains and links?

6. A topographic abney reading of a +36½ was obtained between two points 1.72 chains (horizontal distance) apart. Find the difference in elevation.

7. What slope distance would be measured on a slope on which the topographic abney reading is +65 if the horizontal distance required is 68.0 links? 94.5 links?

8. You wish to locate a road between stations X and Y. Elevation of X = 1010 feet which you set on your aneroid barometer at 9:00 A.M. You then pace up the proposed road to station Y, a horizontal distance of 24 chains and read 1150 feet on the barometer at 10:30 A.M. You return to station X and read 1050 feet on the barometer at 1:30 P.M. What is the difference in elevation between X and Y and what is the minimum grade of a road in percent to connect these two points?

9. The following data was obtained in the field.

Station	S.D.	B.S.	F.S.	Vert. Angle
A				
	140.0 feet	N 70 W		
C				−40%
B				
	180.0 feet		N 60 E	
A				−38%

Find the difference in elevation between B and C.
Find the horizontal distance between B and C.

10. Complete the set of notes below which were obtained in running differential levels between two bench marks. Check the results.

Sta.	B.S.	H.I.	F.S.	Elev.
$B.M._1$	4.62			232.18
TP_1	5.78		5.04	
TP_2	2.23		5.00	
TP_3	6.00		5.83	
TP_4	8.94		4.32	
TP_5	8.04		3.20	
$B.M._2$			3.69	

11. 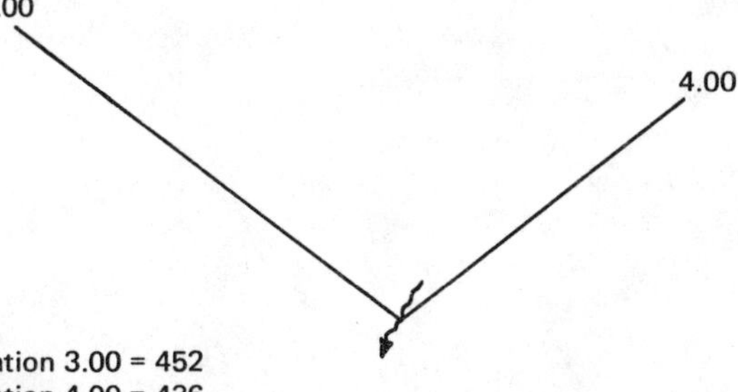

Elevation 3.00 = 452
Elevation 4.00 = 436
Slope distance from 3.00 to creek = 70 links
Abney (topographic) from 3.00 to creek = −55

Find the abney reading from creek to 4.00.
Find the slope distance from creek to 4.00.

12. Explain briefly how you, all alone, would check the adjustment of an abney or clinometer by the two peg method using three nails with no one to place the nails or mark points for you.
13. Find the distance between the 1 chain mark and 26 on the one chain trailer in feet.
14. The survey crew intended set station 8.00 after 6.00 and the head chainman had taken the zero end of the tape ahead to the approximate position of station 8.00. When the head chainman reached this point, the slope had changed so rapidly that it was impossible to see the rear chainman. Hence, station 7.00 would have to be set and the head chainman wanted to set station 7.00 without pulling the tape back so he moved back along the tape toward the rear chainman until he reached the one chain mark. Topographic abneys readings were taken and the slope was found to be a -33 from station 6.00 to 7.00. If the rear chainman held the 2- chain mark on station 6.00, what mark on the tape would the head chainman hold to set station 7.00?
15. Find the slope distance to be measured if the horizontal distance required is 84.5 links and the percent of slope is equal to -64½%

CHAPTER VI. FOREST MAPPING

Section 1. Maps

Forest maps may be classified into two general groups: planimetric and topographic. Planimetric maps show the detail in a flat two-dimensional plane only. By various methods the topographic map shows, in addition to the detail in the flat plane the detail in the third dimension or depth. Contours are the most common type of topographic maps.

Both types of maps show the relative location of objects found on the ground such as buildings, roads, streams, fences, boundaries, etc. Included under boundaries are designations known as "type lines," which denote vegetative cover, ownerships, etc.

Relief may be shown on a topographic map in one of several ways: color, hatch marks, shading, and contours. Maps of large areas such as continents, use different colors to denote each 1000 feet (or other unit) in elevation above mean sea level. Hatch marks are short, straight lines used to denote ridges, peaks, and drainages. The closer the hatch marks are together, the steeper the slope. The same is true in shading where one color is used. Steeper slopes are shaded darker. Contours are considered the most effective method of showing relief and are most commonly used in forest mapping.

Topographic maps are a necessity in planning any work dealing with Forestry. Such maps may be used for:

1. Location of existing features
2. Horizontal distances
3. Elevations and differences in elevation
4. Suitable locations for roads, landings, camps, etc.
5. Logging plans
6. Watershed programs
7. Drainage data and reservoirs
8. Earthwork

Contours

A contour is an imaginary level line on the ground connecting all points of equal elevation. A contour line on a map indicates the elevation above mean sea level. The vertical distance between two adjacent contours is known as the contour interval. The choice of the contour interval depends on the roughness of the terrain, the use which is to be made of the map, and the map scale. The intervals commonly used are 1, 2, 5, 10, 20, 50, and 100 feet. Generally speaking, the larger the scale of the map, the smaller the contour interval. A scale of 1" = 10 feet is considered to be a large scale map, whereas a scale of 1" = 500 feet would be called a small scale map.

Contour lines on the map are similar in shape to contours on the ground but are drawn to the scale of the map. A conception of contours is shown in Figure 36. The upper part of the figure shows two peaks intersected by level and parallel planes spaced uniformly apart. The middle part of the figure shows the areas cut from the solids by the parallel planes. The outlines of the areas are the contours, and these are shown as they would appear on a contour map.

Contours have certain characteristics which are illustrated in Figure 37.

1. All points on any one contour line have the same elevation.
2. Every contour closes on itself either on the map as at A, B, and C or off the map as at F.
3. Summits are indicated by closed contours as at BCDE, and closed contours also indicate depressions (G) which are marked with hatch marks to distinguish from summits.
4. Contours never cross each other except in the case of an overhanding cliff, and then they are dashed lines (H).
5. Contours never split or branch. This is an impossibility in nature as it would indicate a knife-edge ridge.
6. On a plane surface contours would be straight parallel lines.
7. On uniform slopes, the contours are equally spaced as along line ab. Line CD is on a concave slope and EF is a convex slope. The closer the contours are, the steeper the slope.

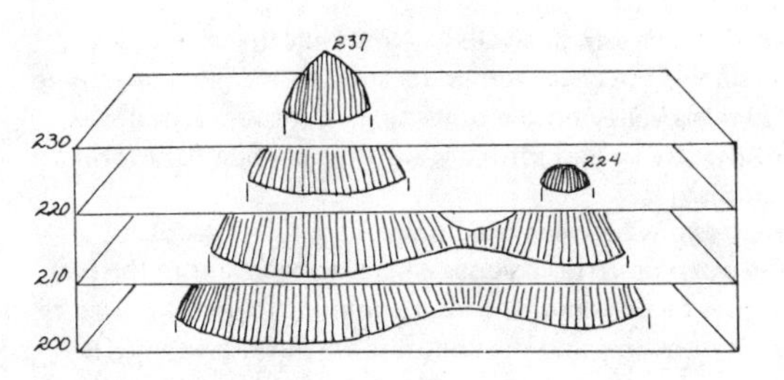

(a) Parallel planes intersecting terrain

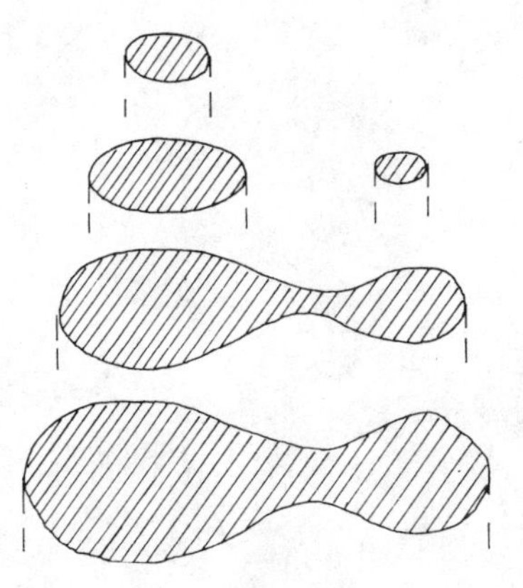

(b) Sections cut by parallel planes

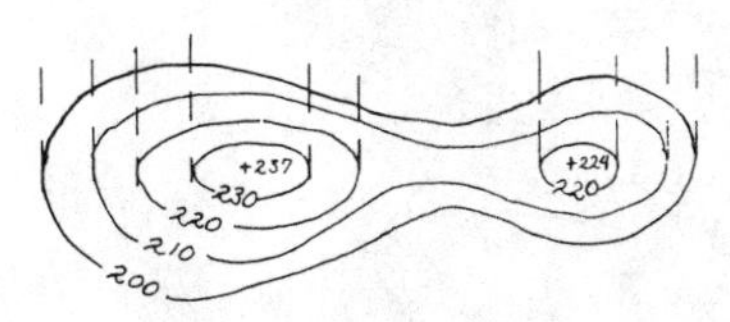

(c) Contour map of terrain in (a)

Figure 36. Contours.

8. Stream JI is a stream in a valley. Note that the contours run up one side of the valley, cross the streams at right angles, and run down the valley on the other side. Contours are convex toward the stream and form a U up stream while they form a U on ridges.
9. The depressions between the summits are called saddles.
10. On very steep or vertical slopes where one contour in the plan view would be on top of the next contour, a symbol is used to indicate a pit or quarry and the contours are not drawn in (K).

Figure 37. Contour map.

Occasionally, to expedite the field work, form lines are used. They have the same characteristics as contours but occur at irregular elevations. They merely indicate the general form of the ground and slope. More office time is required to make a contour map from form lines obtained in the field.

Map Controls

Control in mapping is the skeleton or framework which is surveyed and balanced if need be for purposes of obtaining the correct size, shape, and location of the area being mapped. Controls consist of two components: horizontal and vertical. The former establishes the position in the horizontal plane, and the latter provides the elevations of various points on the horizontal controls.

There are several ways to establish the controls, but only a few are mentioned below beginning with the most accurate. Again, the use of the map prepared would dictate the accuracy desired, and, hence, the method of establishing the controls:

Horizontal Control	Vertical Control	Detail and Relief
Triangulation	Direct levelling	Engineers level and steel tape
Transit traverse	Vertical angles with transit	Hand level or transit and tape
Double abney, compass and tape	Double abney	Strips using compass abney and tape
Single abney, compass and tape	Single abney	Barometer, compass and pacing

The controls are established with instruments of equal accuracy, and the details are usually found with instruments of equal or less accuracy.

Section 2. Field Procedures

The choice of field methods of establishing the controls as previously mentioned will be influenced by the intended use of the map, which in turn will determine the scale and the contour interval which is affected by the size of the area and the topography. Perhaps the most critical item to consider is the allowable budget for such a project.

The first step in mapping is to tie down the controls by locating some known or established point such as a ¼ Section Corner or a Section Corner. This then locates station 0.00, the beginning point in the horizontal plane. This should be done whether the project is a closed traverse or a "P" line for a logging road. The elevation of station 0.00 may be established in one of several ways: (1) assume an elevation, (2) establish an elevation by means of an aneroid barometer, or (3) establish an elevation by differential leveling from a known bench mark.

For explanatory purposes the example to be given will be the map controls for a rectangular area and will be a closed traverse, that is, a traverse which closes on the beginning point. One very important point to remember is that whether the traverse is an open or closed traverse, it is necessary to take foresights and backsights at every compass set-up with the exception of the initial point. This is a must in order to correct for local attraction. The first bearing taken from station 0.00 may be assumed to be correct, or it can be established by a celestial observation.

The detail and relief for the map may be obtained by the strip method (see Figure 38). Since it is easier to map an area by running strips as near perpendicular to the slope as possible, the crew chief must keep this in mind when establishing the controls. Assume that the area which is to be mapped is six chains on a side. Once the bearing of the first course of the horizontal controls has been determined, establish one side of the traverse by running a straight line. This may be done by sighting the entire length of the side using only one set-up of the compass on station 0.00 or by prolonging a straight line by backsighting. Depending on the slope of the ground, stations 1.00, 3.00, and 5.00 may be set on line to serve as control points for the strips.

Figure 38 also shows stations 7.00, 9.00, and 11.00 being set as control points for the strips.

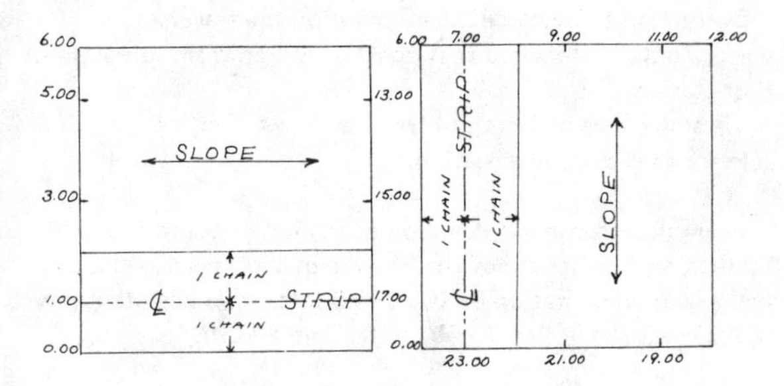

Figure 38. Map controls.

At station 6.00 a backsight is taken to station 0.00, if possible, and a 90° angle turned from the backsight. Since local attraction might exist, the backsight bearing may not agree with the foresight. The bearing must be carefully noted and the new bearing computed from the 90° interior angle, and the line is run from 6.00 to 12.00. At station 12.00, the procedure is repeated. If stations at 1.00, 3.00, and 5.00 were set, then station 13.00, 15.00, and 17.00 would be set between 12.00 and 18.00. If station 7.00, 9.00, and 11.00 were established, then station 19.00, 21.00, and 23.00 would be set between 18.00 and 24.00. These stations will now serve as control points for the strips. The controls are not complete until the traverse has been closed either on the initial point or some other known point.

Three components of closure must be determined if it is a closed traverse: (1) line error, right or left; (2) distance error, long or short; (3) elevation error, high or low. The line and distance error can be adjusted in the process of drawing the map. The elevation error should not exceed one-half of the contour interval. This is only a rule of thumb and will depend mainly on the terrain. If there is still an elevation error after all abney readings and calculations have been double checked, the error is corrected and proportioned around the traverse by a direct proportion method.

$$C = S/T \times E$$

C = Correction to be made at any point on the traverse.
S = Horizontal distance to that point measured from initial point or station.
T = Total distance or length of traverse.
E = Error of closure in elevation.

For example, assume the elevation of the initial point, Sta. 0.00, is 578.0 feet, and the length of the traverse in chains is 40.00 chains. The final elevation of station 0.00 is 590.0 feet. You wish to know the correct elevation of Sta. 13.00, 15.00, and 17.00:

E = 590-578 = +12.0 feet
S = 13.00
T = 40.00

$C_{13} = \frac{13}{40} \times 12 = 3.9$ feet. $C_{15} = \frac{15}{40} \times 12 = 4.5$ feet.

$C_{17} = \frac{17}{40} \times 12 = 5.1$ feet.

The correction which is to be applied takes the opposite sign of the error. Hence if the error is positive, the corrections will be subtracted from the elevations determined and vice versa. If the elevation of station 15.00 was 632.5 then corrected elevation would be 632.5 - 4.5 = 628.0. The stations which are to be the control points for the strips must be corrected before the strips are run to obtain the detail.

Map Detail

The detail for the map is obtained by running strips through the area using a staff compass, topographic tape, and topographic abneys. The control points are the beginning and ending points which are marked with stakes at the intervals previously described. The advantage of strip mapping is that it can be done in the field on the site, and the topographer can do a more accurate job of mapping when he can see the ground slope, etc.

For example: In Figure 39 assume that elevation of a station is 506, and the abney reading to the next station on the center line of the strip is +19. It is necessary to locate the points on the center line of the strip whose elevations are 510 and 520. Assuming the slope to be uniform for one chain, these elevations will therefore be between the points whose elevations are 506 and 525.

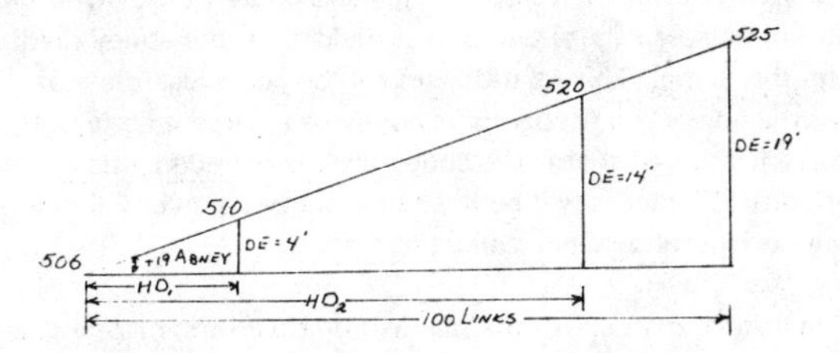

Figure 39. Locating contours in field.

By use of similar triangles then:

$$\frac{HD_1}{100} = \frac{4}{19}$$

HD_1 = Horizontal distance to point whose elevation is 510.
4 = Elevation difference between 506 and 510.
100 = Horizontal distance between two stations.

Hence: $HD_1 = \frac{4}{19} \times 100 = 21$ links or .21 chains.

Similarly:

$$\frac{HD_2}{100} = \frac{14}{19}$$

$HD_2 = \frac{14}{19} \times 100 = 73.69$ links or .74 chains.

This can be simplified by thinking of these values in another way.

4 is the DE.
19 is the abney taken on the slope.
100 links = 1 chain.

Therefore $HD_1 = \frac{DE}{Abney}$ and thus the horizontal distance between any two points on a slope in chains is equal to the difference in elevation between those two points divided by the abney reading taken on this slope. This relationship will be good regardless of whether the abney is a topographic abney or a percent abney. It must be remembered that if the abney used is a topographic abney the horizontal distances will be in chains and if a percent abney is used the horizontal distance will be in feet.

Using a three-man crew, one man would do all the mapping, and the other two crew members would run the compass line and take the abney readings. With the strip beginning at station 9.00 on the controls a backsight from this station along the line from 6.00 to 12.00 and a 90° angle turned with the compass would establish the line of the center line of the strip. An abney reading would be taken at right angles to the center line from station 9.00 along the control lines. Stations would be established at one chain intervals along the center line. Abneys are taken at right angles to the centerline on both sides to determine the location of contours, all other detail would be located by pacing. The mapping is usually done on cross-section paper starting at the bottom of the sheet which is always oriented along the center line. The notes relating to the centerline are kept to one side of the pages so as not to clutter up the detail.

Figure 40, which can be used as a work page for the following example, shows the order for obtaining the detail. (Circled consecutive numbers)

1. Locate the 520 foot contour to the right of ℄ if the abney reading is -22 to the right, and the elevation of 0.00 on ℄ is 522.

   HD = 2/22 x 100 = 9.09 or 9 links.
2. Locate the next contour to the right or the 510 foot contour. It may be located from the ℄ or from the preceding contour.

   From ℄: HD = 12/22 x 100 = 55.5 links.

   From 520' contour HD = 10/22 x 100 = 45.5 links.

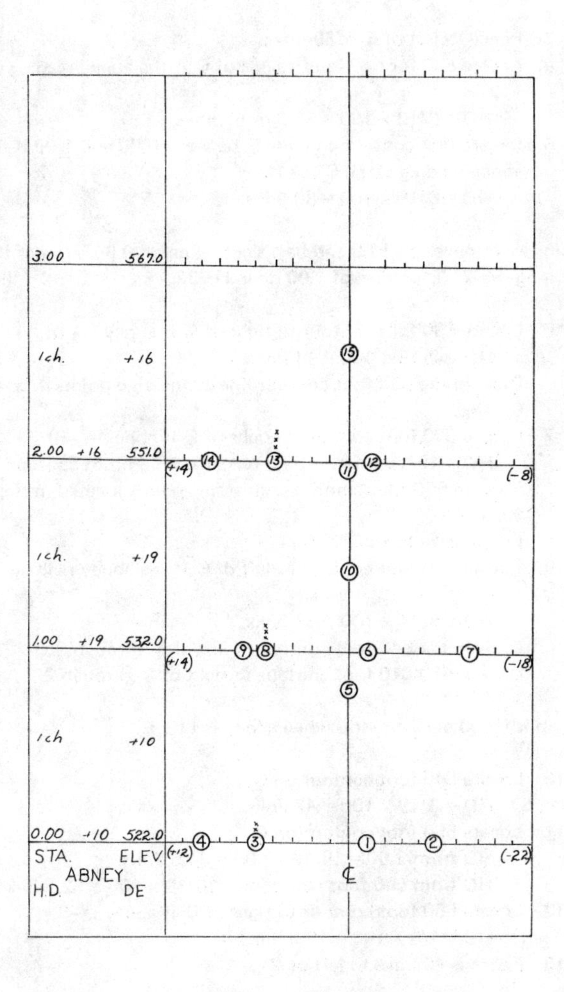

Figure 40. Work sheet for strip mapping.

3. Fence to left of ℄ is 50 links.
4. Locate 530 foot contour to left of ℄ if the abney reading is +12.

   HD = 8/12 x 100 = 66.6 or 67 links.
5. Locate first contour crossing ℄ between 0.00 and 1.00 if the abney reading along ℄ is +10.

   HD = 8/10 x 100 = 80 links.

Compass is moved to 1.00 and backsight taken on 0.00 in order to prolong the ℄. Elevation of 1.00 on ℄ is 532.

6. Locate 530 foot contour to right of ℄ if abney is –18.

   HD = 2/18 x 100 = 11 links.

   Draw in the 530 foot contour line connecting points 4, 5, and 6.
7. Locate 520 foot contour to right of ℄ if abney = –18.

   HD = 10/18 x 100 = 55.5 links from 530 foot contour.

   Draw in 520 foot contour connecting points located in step 1 and 7.
8. Fence 45 links to left of ℄.
9. Locate 540 foot contour to left of ℄ if the abney reading is +14.

   HD = 8/14 x 100 = 57 links.

   Draw in the 540 foot contour to left of ℄ through 9.

   Draw in the 510 foot contour to right of ℄ through 2.

Establish 2.00 on ℄ of strip when abney is +19.

10. Locate 540 foot contour on ℄.

    HD = 8/19 x 100 = 42 links.
11. Locate 550 foot contour on ℄.

    HD from 2.00 = 18/19 x 100 = 94.5 links.

    HD from 540 foot countour = 10/19 x 100 = 52.5 links.
12. Locate 550 foot contour to right of ℄ if abney is –8.

    HD = 1/8 x 100 = 12.5 links.
13. Fence is 40 links to left of ℄.

14. Locate 560 foot contour to left of ℄ if abney is +14.
    HD from 2.00 = 9/14 x 100 = 64.31 links.
    Draw in 550 foot contour through 11 and 12 and off edges of strip.
    Draw in 540 foot contour through 9 and 10 and off edges of strip.

Establish 3.00 on ℄ strip when abney is +16.

15. Locate 560 foot contour on ℄.
    HD = 9/16 x 100 = 56.3 links.
    Draw in 560 foot contour through 14 and 15 and off left edge of strip, etc., etc., etc.

In order to use the strips in preparing the map it is necessary to determine the line error, distance error, and elevation error of closure on each strip, as was done for the map controls. The elevation corrections if needed must be calculated for the closing station ot each strip before the elevation error can be computed.

Line error and distance error on each strip can be adjusted as the map is assembled. However, any elevation error must be corrected before the map is made. This error can be corrected in one of two ways and is done only after checking the field work to assure that the error did not occur all in one place. Thus it may be assumed that the error accumulated gradually.

One method is to correct by proportioning the error over the entire length of the strip by the formula:

$$C = S/TL \times E$$

C = Correction
S = Station of the contour
TL = Length of strip in chains or feet
E = Elevation error of closure.

For example: In running a strip 10 chains in length, the vertical error was found to be a +6 feet. The strip was drawn as follows:

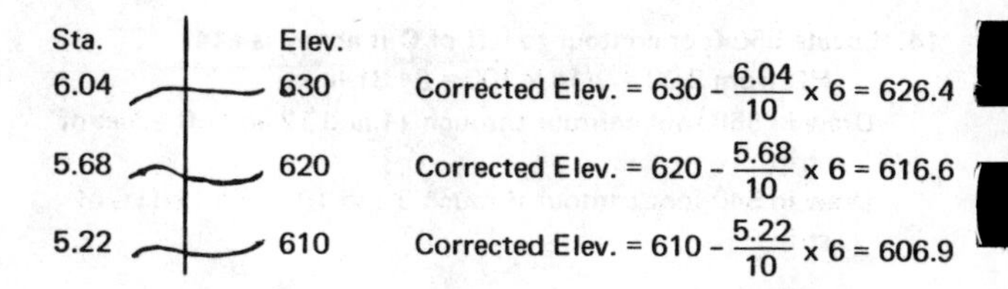

The corrected elevations of stations 604, 568, and 522 are 626.4 feet, 616.6 feet, and 606.9 respectively. Since these elevations do not represent the even 10 ft. contours, it is now necessary to determine the stationing of the 620 ft. contour which obviously lies between the points whose corrected elevations are 616.6 and 626.4. This is done by proportion based on the principal of similar triangles.

$$\frac{HD}{36} = \frac{3.4}{9.8}$$

$$HD = \frac{36 \times 3.4}{9.8} = 12.5 \text{ links}$$

The distance from the original 620 ft. contour to the corrected 620 ft. contour is 12.5 links or the stationing of the latter is 580.5. All of the contours are corrected in this manner, each contour following the form of the original contour adjacent to it.

The second method involves correcting the elevation of each station on the center line, i.e., 1.00, 2.00, 3.00, 4.00, etc. The corrected elevation of each station will provide an adjusted abney reading between each station on the ℄ since the abney reading is equal to the difference in elevation for one chain horizontal distance. Using the adjusted abney reading, relocate each contour between stations on the ℄ as was originally done while mapping in the field.

For example; the elevation error of closure in a strip 10 chains long was -5.5 ft.

Sta.	Elev.	Abney	Corrected Elevation
9.00	489		$9.00 = 489 + \frac{9}{10} \times 5.5 = 493.9$
8.81	485		
8.57	480		
8.33	475		
8.09	470		
8.00	468	+21	$8.00 = 468 + \frac{8 \times 5.5}{10} = 472.4$
7.81	465		
7.50	460		
7.18	455		
7.00	452	+16	$7.00 = 452 + \frac{5.5}{100} \times 7 = 455.8$

℄

Sta.	Elev.	Abney
9.00	493.9	
8.82	490	
8.59	485	
8.35	480	
8.12	475	
8.00	472.4	21.5
7.85	470	
7.55	465	
7.25	460	
7.00	455.8	16.6

℄

The new location of the new contours is based on the newly adjusted abney readings. The new contours are drawn in following the form of the contours adjacent, and then the old contours are erased.

Section 3. Map Assembly

In preparing a finished map, several things may be done before actually drawing the map. To make a completed map, eight items must be included: Title, date, name, contour interval, map scale, legend, datum, and north with proper magnetic declination. The size of the map sheet will be determined by the scale to be used. Figure 41

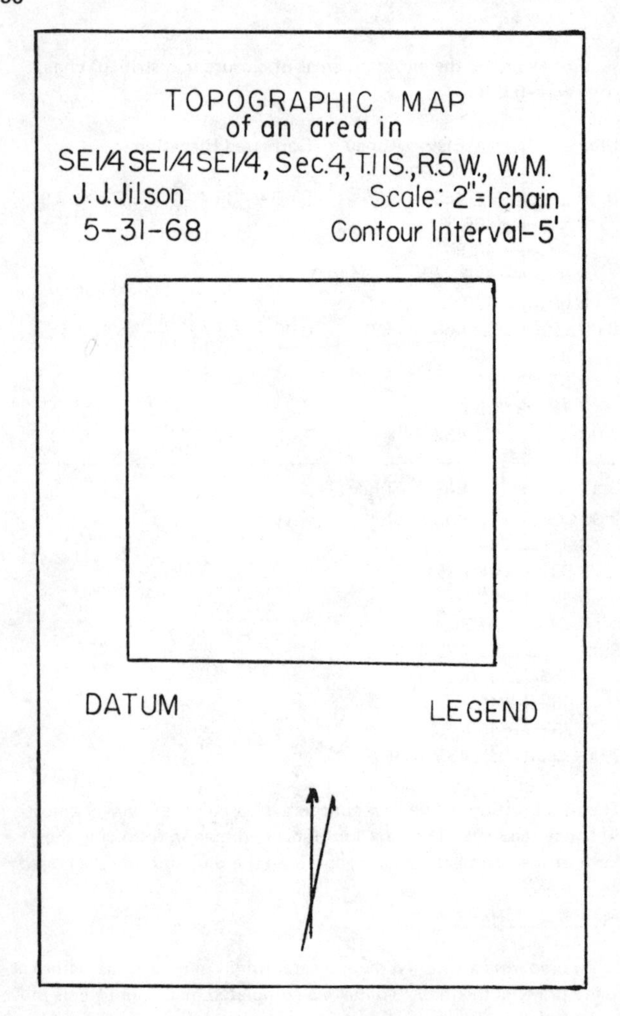

Figure 41. Map format.

shows a proposed arrangement for this information showing an acceptable margin. Careful planning can be made before drawing the map since its size is predetermined by the area mapped and the scale to be used. The lettering size and character will then determine the space needed at the top of the sheet and at the bottom.

The horizontal controls for the map will be shown as a square even though there was some line and distance error occurring in the closing of the traverse. This square can be drawn properly positioned on the sheet.

Since a line and distance error in closing probably exists, it must be taken into account before proceeding with the detail of the area. Assume that there was a line error of 20 links to the left and a distance error of 24 links long. It is then assumed that this error has been gradually accumulated over the entire traverse and should be proportioned around the traverse at the four corners. If it is known that the error is due to improperly turning a corner or taking a poor backsight, the line should be rerun in the field and not corrected as the map is drawn.

Using the above assumed error, the actual traverse run in the field is shown in broken lines and is compared to the closed traverse in solid lines as it should be shown on the finished map in Figure 42.

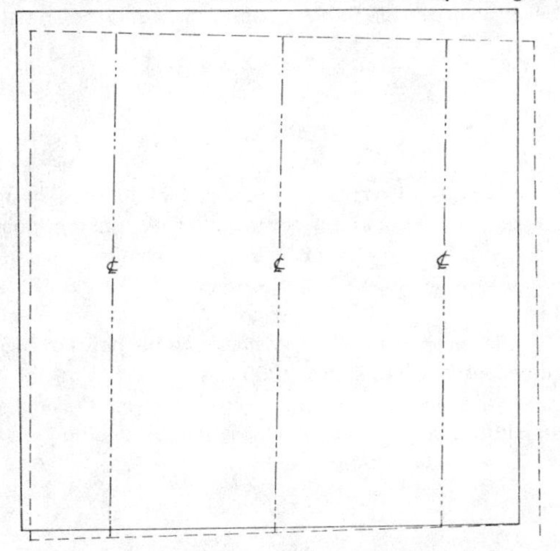

Figure 42. Adapting field traverse to finished form.

The traverse as established in the field is shown very lightly with construction lines and then erased when the map is finished. The stations which served as control points for the strips are then located on the field traverse as shown, and the individual strips are transferred to the map after properly orienting the beginning and ending point according the line and distance error of closure that existed on each strip. The detail can be traced by working on a light table. The detail is finished only after all strip detail has been copied for each strip. Since the contours were approximate at the edge of the strip, they must be united into a smooth contour across the entire map. Contours and other detail outside the closed boundary are omitted but must be extended on other areas so that the finished detail is extended to the edge of the boundary shown on the finished map. As a rule every fifth contour is darker or heavier, and the elevation is shown which can be read from the bottom or the right hand side of the map.

The datum should include the following information: (1) location of station 0.00 in respect to a known corner, (2) equipment with which the controls were established and the detail obtained, (3) length and bearing of each course of the controls, and (4) assumed elevation of station 0.00. A sample datum is shown below.

DATUM

The horizontal and vertical controls were established, using a staff compass, two topographic abneys and a two chain topographic tape, beginning at station 0.00 which is ____ chains ____ and ____ chains ____ of the ____ Corner of Section ____, T. ___ S., R. ___ W., W.M. thence ____ ____ chains, thence ____ ____ chains, thence ____ ____ chains, thence ____ ____ chains to the point of beginning. The assumed elevation of station 0.00 was ___ feet. The detail for the map was obtained by running ___ strips through the area, each two chains in width in a ____ direction using the same equipment with which the controls were established.

The legend consists of symbols, one for each bit of detail that is shown on the map and the most common are shown in Figure 43.

The map is always oriented in such a position that north is toward the top of the paper. The north arrow and the declination of the area is placed between the datum and legend. Proper drafting methods are to be followed in all map preparation.

Section 4. Problems

1. The elevation of station 4.00 on the center line of the strip is 647.5 feet. The topographic abney reading to station 5.00 on the center line is +32. At what station does the 660 foot contour cross the center line of the strip?

2. The length of map strip = 1320 feet. The elevation error in closing = –12 feet.

   Uncorrected station of 840 foot contour = 8 + 16
   Uncorrected station of 820 foot contour = 7 + 43
   Uncorrected station of 810 foot contour = 6 + 15

   What is the corrected station of the 820 foot contour?

3. What is the horizontal distance between 10 foot contours with the following abneys?

   Topographic abney = +28½
   Percent abney = +36
   Degree abney = $+17^{o}00'$

4. The elevation of 0.00 = 840 feet and the elevation of station 40.00 = 852 feet where 0.00 is the initial point and 40.00 is the last station of the map controls of an area 10 chains on a side.

   The line error was 36 links right, the distance error was 24 links long and the topographic abney reading from 40.00 to 0.00 was = –20. What is the elevation error for this traverse?

5. In the above problem what correction in elevation should be made at station 17.00 and what is the sign of the correction?

6. The elevation of stations 4.00 and 5.00 on a map strip which is 6 chains in length are 1021.5 and 1043.5 respectively. The elevation error on this strip was a +8 feet. What is the corrected station of the 1025 foot contour?

7. Fill in the blanks in the note form below.

Sta.	H.D.	Abney (topog.)	D.E.	Elev.
———				848
	1.40		———	
———		———		———
	3.00		———	
3.00		-10		862
	———		———	
2.00		———		———
	2.00		———	
0.00		+20		840

8. On a map whose scale is 1" = 400', the distance between two adjacent contours is 0.26 inches and the contour interval is 50 feet. What is the percent of slope? On this same map, what would the distance be between contours if the slope were 30%?

9. If the difference in elevation between two point 0.38 chains apart (H.D.) is 20 feet, what topographic abney reading would you obtain on this slope?

10. What is the area in acres that can be logged off around a 160.0 foot spar pole set on a uniform 68% slope if the length of line available is 850 feet? (Line length is from top of spar pole to the ground.)

11. The distance from a mark on the one chain trailer to a like mark on the two chain trailer equals 122.75 links. What is the mark?

12.

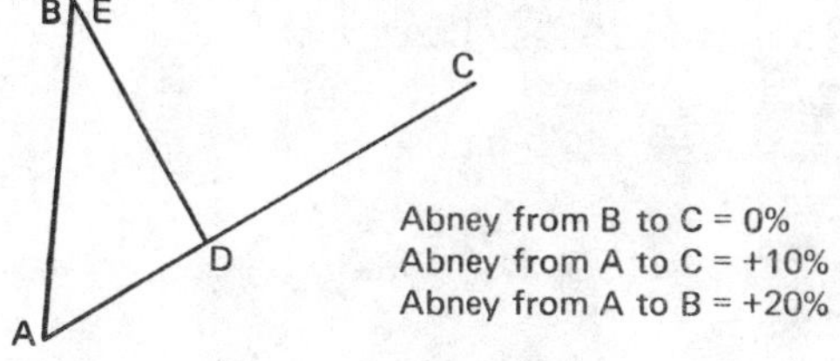

Find the precent abney reaching from D to E. (D may be any place on AC but assume that B & E coincide.)

13. The ground spacing of coutours (horizontal distance between contours) may be obtained by multiplying the contour interval by the natural _____ function of the angle of the slope.

14. Find the distance in feet from 12 on the one chain trailer to 29 on the two chain trailer.

15. Compute the slope you would read on the topographic and percent scales respectively on a clinometer if the slope distance between adjacent contours equals 77 feet and the contour interval is equal to 25 feet.

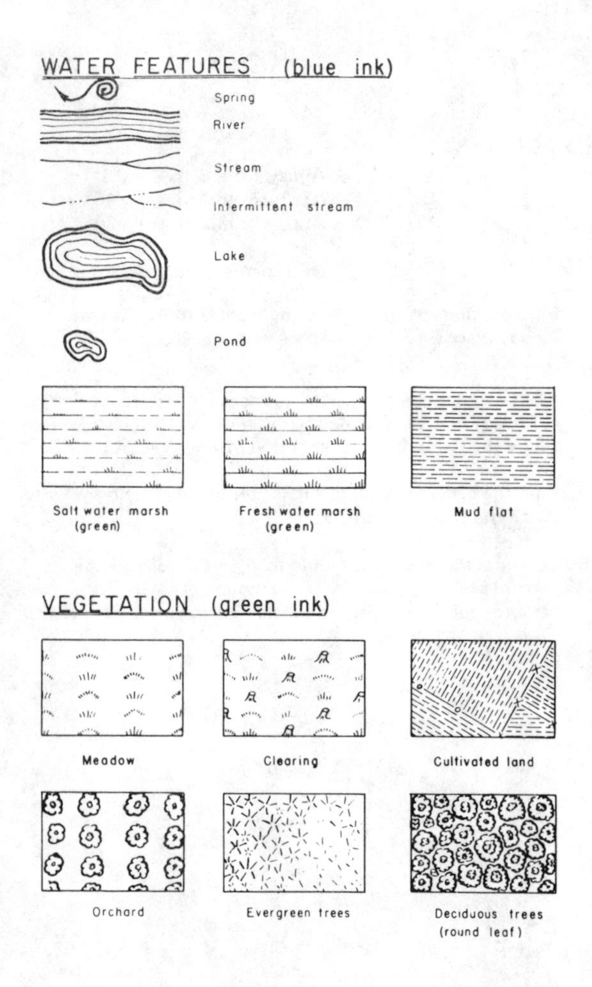

Figure 43. Conventional signs and symbols.

CULTURE & WORKS of MAN (black ink)

Township line

Section line

$\frac{1}{4}$ – $\frac{1}{16}$ Section line

Survey corner located

" " not located

Triangulation station

BM
X
1510

Permanent benchmark & elevation

Main motor road (paved or macadamized)

Motor road, good (gravel – all weather)

" " , poor (summer only)

Pack trail, good

" " , poor

Railroad track, double line

" " , single "

" switch

Telephone line

" " along road

" " " trail

Wire fence, barbed

" " , smooth

Mine or quarry

Building, occupied

" , unoccupied

Bridge

Dam

RELIEF FEATURES (brown)

600

500

Contours

320

340

Depression contours

Figure 43. (Continued).

APPENDIX

SLIDE RULE

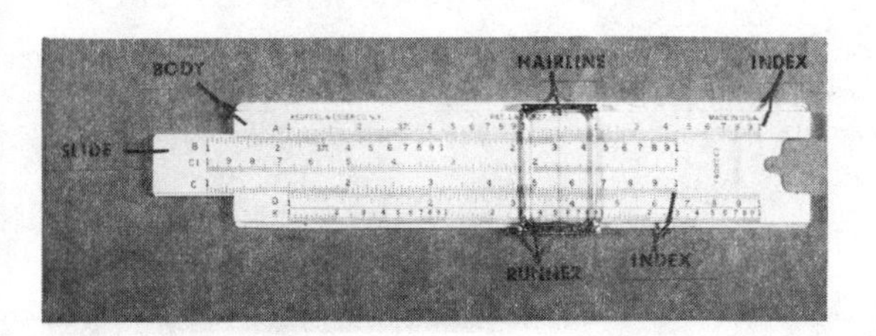

To use a slide rule efficiently with a minimum of error requires a certain amount of practice. An examination of the D scale will show that it is divided into 9 parts by primary numbers which are numbered 1, 2, 3,...,9,1. The space between any two primary marks is divided into ten parts by nine secondary marks. These are not numbered on the actual scale except between the primary marks numbered 1 and 2.

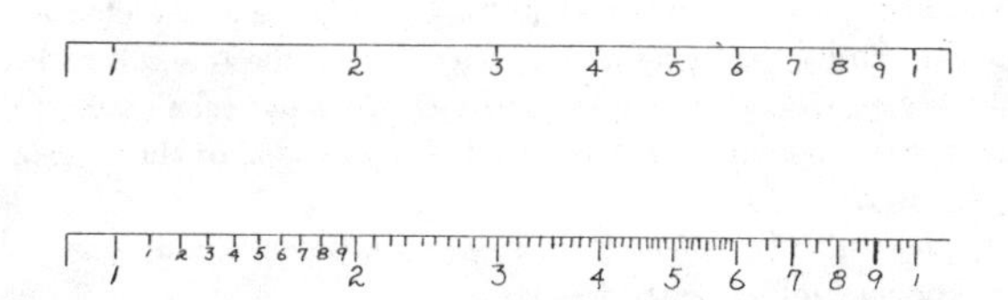

Between the secondary marks appear smaller or tertiary marks which aid in obtaining the third digit of a reading. Thus, between the secondary marks numbered 23 and 24 are 4 tertiary marks. If you think of the end marks as representing 230 and 240, the four tertiary marks divide the intervals into five parts, and with these marks we associate the numbers 232, 234, 236, and 238. The reading of any position between successive tertiary marks must be based on an estimate.

All of the spacing of primary, secondary, and tertiary marks from left to right is determined by the logarithm of these numbers and these logarithmic spacings make possible the various calculations that can be performed on a slide rule.

It is important to note that the decimal point has no bearing upon the position associated with the number on the C and D scales. If the hair line were set as shown the principal digits would be 2, 2, 6

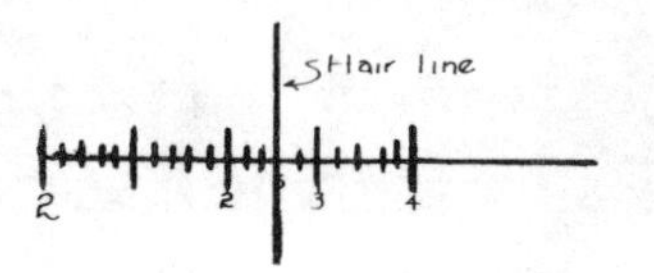

but the number may represent 226, 2.26, 0.00226, 2,260, or 22.6.

Assuming that the error of reading is one tenth of the smallest interval following the left hand index of D, the error is roughly 1 in 1000 or one tenth of one percent on a 10-inch slide rule.

Multiplication

Multiplication is performed on the C and D scales, which are identical. Multiplication may be performed on the A and B scales, which are also identical, but since the divisions between marks are smaller, the accuracy will be reduced. Either index of the C scale may be used.

1. Set the index of the C scale opposite one of the number to be multiplied on the D scale.
2. Move the hairline to the other number of the C scale.
3. Read answer under hairline on the D scale.
4. The decimal point is placed in accordance with the result of a rough calculation. Perform the indicated multiplications.

1. 3 x 5
2. 5.56 x 634
3. 0.0495 x 0.0267
4. 1.876 x 5.32
5. 912 x 0.267
6. 0.298 x 0.544
7. 48.7 x 1.173
8. 743 x 0.0567
9. $(75.0)^2$
10. 4.98 x 576

To multiply three or more factors, multiply the first two in the usual way. The hairline is now over the first product on the D scale. Move the index of the C scale to the hairline and then move the hairline to the third factor on the C scale. The final product is under the hairline on the D scale.

Division

1. Set the divisor on the C scale directly over the dividend on the D scale.
2. Move the hairline to the index on the C scale.
3. Read the quotient under the hairline on the D scale.
4. The position of the decimal point is determined from rough calculation. Perform the indicated operations.

1. 3.75/0.0227
2. 1029/9.70
3. 2875/37.1
4. 0.0385/0.001462
5. 3.42/272
6. 0.0456/9.9297
7. 0.0592/1.983
8. 10.05/30.3
9. 87.5/37.7
10. 871/0.468

Like the C and D scale, the CI is a single unit logarithmic scale. The CI scale is an inverted C scale—its numbers increase from right to left. Because of this arrangement, the values of the CI and C scales are the reciprocals of each other. To multiply using the CI scale, set one number on the CI scale opposite the other number on the D scale. Read the answer on the D scale directly below the index of the C scale. In division using the CI scale, set the index of the CI scale over the dividend on the D scale. Move the hairline to the divisor on the CI scale, and read the answer under the hairline on the D scale.

In solving problems involving both multiplication and division:

1. Divide the first number in the numerator by first number in the denominator.
2. Multiply this result by second number in the numerator (switching use of C index if necessary).
3. Divide this result by second number in the denominator.

4. Continue alternately multiplying and dividing until all indicated operations have been completed.

$$\frac{54 \times 68 \times 132 \times 81/6}{72 \times 16 \times 43 \times 21} = 6.29 \qquad \frac{2.82 \times 6.95 \times 7.85 \times 436}{79.8 \times 0.0317 \times 870} = 30.5$$

Squares and Square Root

To solve problems involving squares and square roots, only the A and D scales are used, and there is no need for the slide of the slide rule.

Squares

1. Set hairline over the number on the D scale.
2. Read its square under the hairline on the A scale.

The square roots are found on the slide rule by the exact opposite method from that used in finding squares. The hairline is set over the number on the A scale and the square root is read under the hairline on the D scale. However, since the A scale is only one half the D scale, i.e., is a two-section scale with two sets of numbers 1 through 10, it is necessary to determine by the following method which half of the A scale to use.

If the number is 1 or larger.

1. Space off groups of two digits, starting at decimal point and working to the left.
2. If last group to left has one digit, use left half of the A scale.
3. If the last group to left has two digits, use the right half of the A scale.

If the number is less than 1.

1. Space off groups of two digits, starting at decimal point and working to right.
2. If the first group with significant figures to the right of the decimal point is a one-place number, use left half of A scale.

3. If a two-place number, use right half of A scale.

1. $42.2\sqrt{0.328}$
2. $1.83\sqrt{0.0517}$
3. $\sqrt{51.7} \div 103$
4. $\frac{5.66 \times (7.48)^2}{79}$
5. $\frac{(2.38)^2 \times 19.7}{18.14}$
6. $20.6 \times \sqrt{7.89} \times \sqrt{1.571}$

Cubes and Cube Roots

To solve problems involving cubes and cube roots only the D and K scales are used.

Cubes

1. Place hairline over the number to be cubed on the D scale.
2. Read the cube of the number under the hairline on the K scale.

The cube roots are found on the slide by the exact opposite method of that used in finding cubes. The hairline is set over the number on the K scale, and the cube root is read under the hairline on the D scale. However, it is necessary to determine which section of the K scale should be used since there are three sections, each only one third of the D scale, with three sets of numbers from 1 to 10.

If the number of which the cube root is to be found is 1 or larger:

1. Space off groups of three digits, starting at the decimal point and work to the left.
2. If the last group to the left has one digit, use the left hand section of the K scale.
3. If two digits, use the middle section.
4. If three digits, use the right hand section.

If the number is less than 1:

1. Space off groups of three digits, starting at decimal point and work to the right.

2. If the first group with a significant number to right of decimal point is a one-place number, use the left hand section of the K scale.
3. If a two-place number, use the middle section of the K scale.
4. If a three-place number, use the right hand section of the K scale.

1. $\sqrt[3]{73.2 \times 0.523}$
2. $\sqrt[3]{9.72}$
3. $(21.3)^3$
4. $(72.3)^2 \times 8.25$
5. $27\pi \div \sqrt[3]{661{,}000}$
6. $\dfrac{\sqrt[3]{32.1 \times 0.0585 \times \pi}}{1/3.63}$
7. $489 \div \sqrt[3]{732}$
8. $3.83 \times 6.26 \times \sqrt[3]{54.2}$
9. $\sqrt[3]{97.2}$
10. $(.928)^3$

Logarithms

The fractional part of a logarithm is called the mantissa, and the number preceding the decimal point is called the characteristic. For 1 and numbers greater than 1, the characteristic is one less than the number of digits to the left of the decimal point. For numbers less than 1, the characteristic is negative and is one more than the number of zeros to the right of the decimal point.

To find the log of a number, set the hairline over the number on the D scale and read the mantissa under the hairline on the L scale. Determine the characteristic as described above.

1. 32.7
2. 6.51
3. 0.676
4. 0.01052
5. 72.6
6. 0.267
7. 0.00802
8. 432

Trigonometry

Trigonometric operations are those involving the ratios: sine, cosine, tangent, cosecant, secant and cotangent of angles. These angular functions are the ratios of the lengths of the sides of a right triangle.

The sine or tangent of angles smaller than 0.57 degrees can be easily obtained by use of the approximate relation:
Sine 0° = Tangent 0° = 0° (radians) approximately. Hence, $1^o = \pi/180$ radians, $1' = \pi/180 \times 60$ radians and $1'' = \pi/180 \times 60 \times 60$ radians for small angles. For convenience, the value of $180 \times 60/\pi$ has been marked by a "minute" gauge point near the 2 degree division and the value of $180 \times 60 \times 60/\pi$ has been marked by a "second" gauge point near the 1.167 division. To find the value of a small angle set the proper gauge point opposite the number on the D scale and read the value opposite the index of the C scale on the D scale. The number of zeros following the decimal point is determined accordingly for the sine of the angle.

	Angle	Sine
Between	0° 00′ 00″ & 0° 00′ 21″	0.0000-
	0° 00′ 22″ & 0° 03′ 24″	0.000--
	0° 03′ 25″ & 0° 34′ 24″	0.00---
	0° 34′ 25″ & 5° 44′ 20″	0.0----
	5° 44′ 21″ & 90° 00′ 00″	0.-----

The number of zeros following the decimal point in the tangent of the angle is the same except that there is no zero just to the right of the decimal at 5° 42′ 38″ and the tangent exceeds 1.0000 after the angle exceeds 45° 00′ 00″.

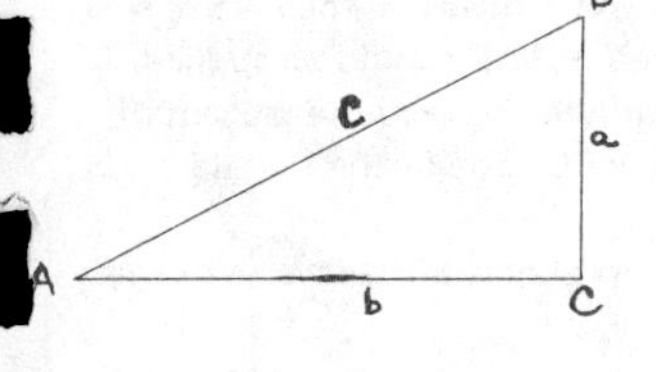

sin A = a/c   cosec A = c/a = 1/sin A

cos A = b/c   sec A = c/d = 1/cos A

tan A = a/b   cot A = b/a = 1/tan A

sin A = cos (90-A)   cos A = sin (90-A)   tan A = cot (90-A)

cot A = tan (90-A)   versine A = 1 - cos A   exsec A = $\frac{1-\cos A}{\cos A}$

The sine varies from 0.10 to 1.0 for angles varying from 5.74 to 90 degrees, whereas the cosine varies from 1.0 to 0.10 for angles varying from 0 to 84.3 degrees. The sine and cosine functions are determined from the S scale. The tangent varies from 0.10 to 1.0 for angles varying from 5.71 to 45 degrees, and the tangent varies from 1.0 to 10.0 for angles between 45 and 84.29 degrees. Sines and tangents of angles between 0.57 and 5.73 degrees are identical and will vary from 0.01 to 0.10. For angles of this size use the ST scale.

To find the sine function of an angle, line up the indexes of the S and D scales, set the hairline over the angle on the S scale, and read the natural sine on the D scale. If the sine and tangent scale are on the reverse side of the slide, the sine function will be found under the right or left index of the A scale. Other types of slide rules not having a ST scale will read the sine function under the hairline on the B scale.

To find the tangent function of an angle up to 45 degrees, line up the indexes of the S and D scales, set the hairline over the angle, and read the natural tangent on the D scale. If the sine and tangent function are on the reverse side of the slide, the tangent function will be found above the right or left index of the D scale. Other types of slide rules not having a ST scale will read the tangent function under the hairline on the C scale.

The tangent function of any angle greater than 45 degrees will be read on the CI scale after setting the hairline over the angle on the T Scale. For slide rules having the S and T scales on the reverse side of the slide, the formula, $\tan A = 1/\tan(90-A)$ should be applied. Set the value of $(90-A)$ under the hairline and read the tangent of the angle under the right or left Index of C scale on the D scale.

To find an angle when the function is known, the reverse of the above procedure is followed.

1. $\sin 76^\circ$
2. $\sin 54^\circ 30'$
3. $\cos 34.5^\circ$
4. $\cos 74.7^\circ$
5. $\sin 3^\circ 51' 36''$ (3.86°)
6. $\tan 15^\circ 42'$
7. $\tan 49^\circ 18'$
8. $\tan 77^\circ 30'$
9. $\tan 2^\circ 24'$
10. $\sin 0^\circ 54'$ (0.9°)
11. 22°
12. 38.3°
13. 4.7°
14. 38.15°
15. 77.91°

In multiplication, division, and proportion with sines and tangents use the proper scale and proceed as you would in multiplying or dividing two numbers on the C and D scales.

1. $\dfrac{18.6 \sin 36^{o}}{\sin 21^{o}}$

2. $\dfrac{4.2 \tan 38^{o}}{\sin 45.5^{o}}$

3. $\dfrac{13.1 \cos 40^{o}}{\tan 35.2^{o}}$

4. $\dfrac{1 \sin 22.7^{o}}{\tan 28.2^{o}}$

5. $\dfrac{\sin 51.5^{o}}{(39.1)(6.28)}$

6. $\dfrac{\sqrt[3]{6.1\ (4.91)}}{\tan 13.23^{o}}$

The solution of triangles can be found by applying one of the three following laws.

Law of Sines: $a/\sin A = b/\sin B = c/\sin C$

Law of Tangents: $\tan \frac{1}{2}(A-B) = \dfrac{a-b}{a+b} \tan \frac{1}{2}(A+B)$

Law of Cosines: $\cos A = \dfrac{b^2+c^2-a^2}{2bc}$; $a = \pm\sqrt{b^2+c^2-2bc \cos A}$

Also $\sin A = 2\,\dfrac{\sqrt{s(s-a)(s-b)(s-c)}}{bc}$ where $s = \frac{1}{2}(a+b+c)$

Solve the following triangles:

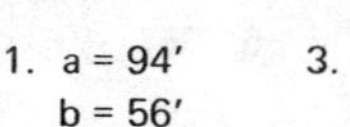

1. a = 94′
   b = 56′
   C = 29°

2. a = 100′
   c = 130′
   B = 51.8°

3. a = 111′
   b = 145′
   c = 40′

4. a = 18′
   b = 20′
   A = 55.4°

5.

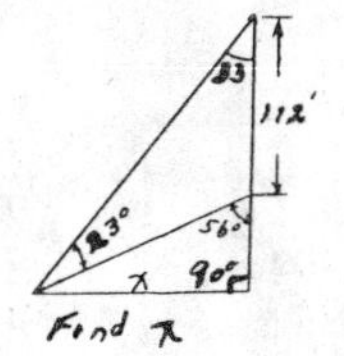

6.

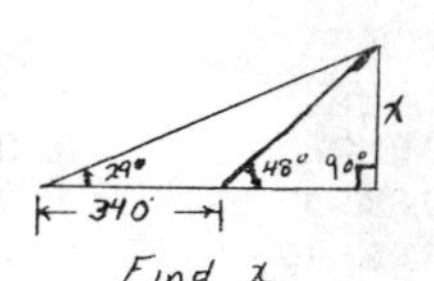

Answers

Multiplication

1. 15
2. 3525
3. 0.001322
4. 9.98
5. 243.5
6. 0.1621
7. 57.1
8. 42.1
9. 5630
10. 2870

Division

1. 165.1
2. 106.1
3. 77.5
4. 26.35
5. 0.01258
6. 0.00469
7. 0.02985
8. 0.3315
9. 2.32
10. 1861

Squares

1. 24.2
2. 0.416
3. 0.0698
4. 4.01
5. 6.14
6. 72.25

Cubes

1. 3.37
2. 2.13
3. 9660
4. 43,100
5. 0.974
6. 2.12
7. 54.3
8. 90.7
9. 4.6
10. 0.80

Logarithms

1. 1.515
2. 0.814
3. 9.830–10
4. 8.022–10
5. 1.861
6. 9.427–10
7. 7.904–10
8. 2.635

Triangles

1. $A = 119.9^\circ$
   $B = 31.1^\circ$
   $c = 52.6'$

2. $A = 49.05^\circ$
   $C = 79.15^\circ$
   $b = 104'$

3. $A = 27.35^\circ$
   $B = 143.1^\circ$
   $C = 9.55^\circ$

4. $B_1 = 66.1^\circ$
   $C_1 = 58.5^\circ$
   $c_1 = 18.6'$
   $B_2 = 113.9^\circ$
   $C_2 = 10.7^\circ$
   $c_2 = 4.08'$

5. 129.4

6. 367

Trigonometric Function

1. 0.970
2. 0.814
3. 0.824
4. 0.264
5. 0.0673
6. 0.281
7. 1.163
8. 4.51
9. 0.0419
10. 0.0157

Find the Angle

11. sin – = 0.375
12. sin – = 0.62
13. tan – = 0.082
14. tan – = 0.785
15. tan – = 4.67

1. 30.5
2. 4.6
3. 14.23
4. 0.720
5. 0.00319
6. 38.15

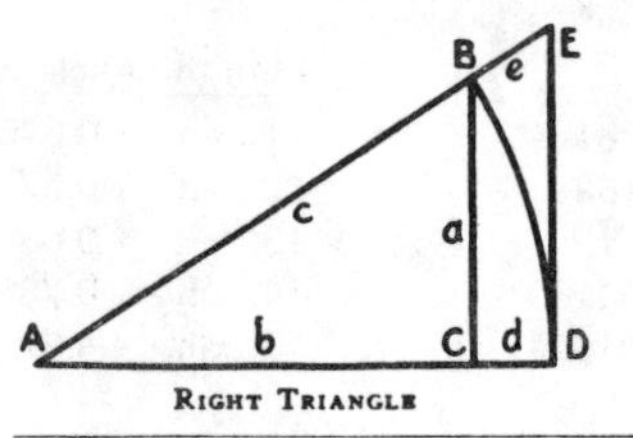

Right Triangle

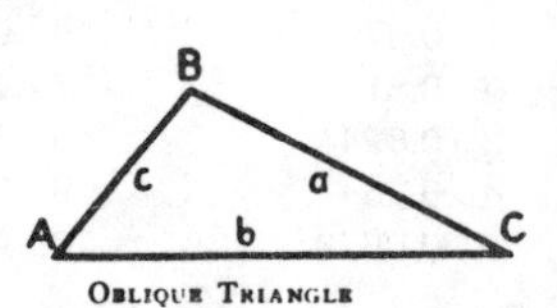

Oblique Triangle

SOLUTION OF RIGHT TRIANGLES.

$\sin A = \frac{a}{c} = \cos B$ $\cos A = \frac{b}{c} = \sin B$

$\tan A = \frac{a}{b} = \cot B$ $\cot A = \frac{b}{a} = \tan B$

$\sec A = \frac{c}{b} = \operatorname{cosec} B$ $\operatorname{cosec} A = \frac{c}{a} = \sec B$

$\operatorname{vers} A = \frac{c-b}{c} = \frac{d}{c}$ $\operatorname{exsec} A = \frac{e}{c}$

$a = c \sin A = b \tan A = c \cos B = b \cot B = \sqrt{(c+b)(c-b)}$

$b = c \cos A = a \cot A = c \sin B = a \tan B = \sqrt{(c+a)(c-a)} = c - c \operatorname{vers} A$

$d = c \operatorname{vers} A$ $e = c \operatorname{exsec} A$

$c = \frac{a}{\cos B} = \frac{b}{\sin B} = \frac{a}{\sin A} = \frac{b}{\cos A} = \frac{d}{\operatorname{vers} A} = \frac{e}{\operatorname{exsec} A} = b + b \operatorname{exsec} A$

SOLUTION OF OBLIQUE TRIANGLES.

Given.	Sought.	Formulas.
A, B, a	b, c	$b = \frac{a}{\sin A} \cdot \sin B,$ $c = \frac{a}{\sin A} \sin(A+B)$
A, a, b	B, c	$\sin B = \frac{\sin A}{a} \cdot b,$ $c = \frac{a}{\sin A} \cdot \sin C.$
C, a, b	$A-B$	$\tan \frac{1}{2}(A-B) = \frac{a-b}{a+b} \tan \frac{1}{2}(A+B)$
a, b, c	A	If $s = \frac{1}{2}(a+b+c)$, $\sin \frac{1}{2} A = \sqrt{\frac{(s-b)(s-c)}{bc}}$
		$\cos \frac{1}{2} A = \sqrt{\frac{s(s-a)}{bc}}$; $\tan \frac{1}{2} A = \sqrt{\frac{(s-b)(s-c)}{s(s-a)}}$
		$\sin A = \frac{2\sqrt{s(s-a)(s-b)(s-c)}}{bc}$;
		$\operatorname{vers} A = \frac{2(s-b)(s-c)}{bc}$
	area	$\text{area} = \sqrt{s(s-a)(s-b)(s-c)}$
A, B, C, a	area	$\text{area} = \frac{a^2 \sin B \sin C}{2 \sin A}$
C, a, b	area	$\text{area} = \frac{1}{2} a b \sin C.$

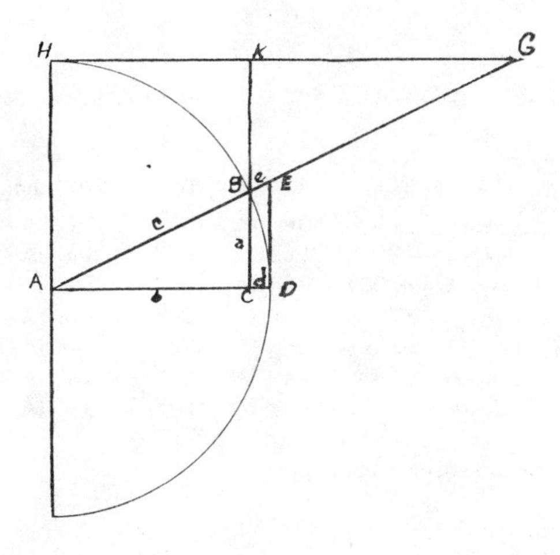

c = 100′

Angle BAC = 30°

$\text{Sin } A = \frac{a}{c}$; $a = (\text{Sin } 30^{o})(100) = 50.00'$

$\text{Cos } A = \frac{b}{c}$; $b = (\cos 30^{o})(100) = 86.60'$

$\text{Tan } A = \frac{a}{b} = \frac{DE}{c}$; $DE = (\tan 30^{o})(100) = 57.70'$

$\text{Cotan } A = \frac{b}{a} = \frac{HG}{c}$; $HG = (\text{cotan } 30^{o})(100) = 173.21'$

$\text{Cosec } A = \frac{c}{a} = \frac{AG}{c}$; $AG = (\text{cosec } 30^{o})(100) = 200.00'$

$\text{Sec. } A = \frac{c}{b} = \frac{AE}{c}$; $AE = (\text{Sec. } 30^{o})(100) = 115.47'$

$\text{Versin } A = \frac{CD}{c}$; $CD = (\text{Vers } 30^{o})(100) = 13.40'$; $\text{versin } A = (1 - \cos A)$

$\text{Exsec } A = \frac{BE}{c}$; $BE = (\text{Exsec } 30^{o})(100) = 15.47'$; $(\text{exsec } A = \frac{1 - \cos A}{\cos A})$

$\text{Tan } \tfrac{1}{2}A = \frac{CD}{BC} = \frac{\text{vers } A}{\sin A} = \frac{d}{a}$

Functions of Angles Greater Than 90° and Less Than 360°

When angle A is between	90° and 180°	180° and 270°	270° and 360°
Sin A	+ Cos (A – 90°)	– Sin (A – 180°)	– Cos (A – 270°)
Cos A	– Sin (A – 90°)	– Cos (A – 180°)	+ Sin (A – 270°)
Tan A	– Cot (A – 90°)	+ Tan (A – 180°)	– Cot (A – 270°)
Cot A	– Tan (A – 90°)	– Cot (A – 180°)	– Tan (A – 270°)
Sec A	– Scs (A – 90°)	– Sec (A – 180°)	+ Csc (A – 270°)
Csc A	+ Sec (A – 90°)	– Csc (A – 180°)	– Sec (A – 270°)
Vers A	1 ÷ + Sin (A – 90°)	1 ÷ + Cos (A – 180°)	1 ÷ – Sin (A – 270°)

Functions of Complements and Supplements

$$\text{Sin A} = \sin(180^\circ + A) = \sin(180^\circ - A) = \sin(A - 180^\circ)$$
$$= \cos(90^\circ + A) = \cos(90^\circ - A) = \cos(A - 90^\circ)$$
$$\text{Cos A} = \cos(180^\circ + A) = \cos(180^\circ - A) = \cos(A - 180^\circ)$$
$$= \sin(90^\circ + A) = \sin(90^\circ - A) = \sin(A - 90^\circ)$$
$$\text{Tan A} = \tan(180^\circ + A) = \tan(180^\circ - A) = \tan(A - 180^\circ)$$
$$= \cot(90^\circ + A) = \cot(90^\circ - A) = \cot(A - 90^\circ)$$

The sign must be determined in each case by considering the quadrant in which the angle lies.

Quadrants	1	2	3	4
Angles	0–90°	90°–180°	180°–270°	270°–360°
Sine and Cosecant	+	+	–	–
Tangent & Cotangent	+	–	+	–
Cosine and Secant	+	–	–	+

Table 1. Field table for percent abney

%	100x Vers	100x Sine	100x Exsec	10x Cotan	%	100x Vers	100x Sine	100x Exsec	10x Cotan
0	0.00	0.0	.00	----	46	9.15	41.8	10.07	21.7
1	0.00	1.0	.00	1000.0	47	9.49	42.5	10.49	21.3
2	0.02	2.0	.02	500.0	48	9.87	43.3	10.95	20.8
3	0.04	3.0	.04	333.3	49	10.20	44.0	11.36	20.4
4	0.08	4.0	.08	250.0	50	10.56	44.7	11.80	20.0
5	0.12	5.0	.12	200.0	51	10.92	45.4	12.25	19.6
6	0.18	6.0	.18	166.7	52	11.28	66.1	12.71	19.2
7	0.24	7.0	.24	142.9	53	11.64	46.8	13.18	18.9
8	0.32	8.0	.32	125.0	54	12.02	47.5	13.67	18.5
9	0.40	9.0	.40	111.1	55	12.38	48.2	14.13	18.2
10	0.50	10.0	.50	100.0	56	12.75	48.9	14.61	17.9
11	0.60	10.9	.61	90.9	57	13.12	49.5	15.10	17.5
12	0.71	11.9	.72	83.3	58	13.49	50.2	15.59	17.2
13	0.83	12.9	.84	76.9	59	13.87	50.8	16.11	17.0
14	0.96	13.9	.97	71.4	60	14.25	51.5	16.62	16.7
15	1.11	14.8	1.12	66.7	61	14.63	52.1	17.14	16.4
16	1.26	15.8	1.27	62.5	62	15.01	52.7	17.66	16.1
17	1.41	16.8	1.43	58.8	63	15.39	53.3	18.19	15.9
18	1.58	17.7	1.60	55.6	64	15.77	53.9	18.72	15.6
19	1.76	18.7	1.79	52.6	65	16.16	54.5	19.27	15.4
20	1.94	19.6	1.98	50.0	66	16.54	55.1	19.82	15.2
21	2.14	20.6	2.18	47.6	67	16.92	55.7	20.37	14.9
22	2.34	21.5	2.39	45.5	68	17.31	56.2	20.93	14.7
23	2.54	22.4	2.61	43.5	69	17.69	56.8	21.49	14.5
24	2.76	23.3	2.81	41.7	70	18.08	57.4	22.07	14.3
25	2.98	24.2	3.08	40.0	71	18.46	57.9	22.64	14.1
26	3.22	25.2	3.32	38.5	72	18.85	58.4	23.22	13.9
27	3.46	26.1	3.58	37.0	73	19.23	59.0	23.81	13.7
28	3.70	27.0	3.84	35.7	74	19.61	59.5	24.40	13.5
29	3.96	27.8	4.12	34.5	75	20.00	60.0	25.00	13.3
30	4.22	28.7	4.40	33.3	76	20.40	60.05	25.63	13.2
31	4.48	29.6	4.69	32.3	77	20.77	61.0	26.21	13.0
32	4.76	30.5	4.99	31.2	78	21.15	61.5	26.82	12.8
33	5.04	31.3	5.30	30.3	79	21.53	62.0	27.44	12.7
34	5.32	32.2	5.62	29.4	80	21.93	62.5	28.06	12.5
35	5.61	33.0	5.95	28.6	81	22.29	62.9	28.93	12.3
36	5.91	33.9	6.28	27.8	82	22.67	63.4	29.32	12.2
37	6.21	34.7	6.62	27.0	83	23.05	63.9	29.96	12.0
38	6.52	35.5	6.98	26.3	84	23.43	64.3	30.60	11.9
39	6.83	36.3	7.34	25.6	85	23.81	64.8	31.24	11.8
40	7.15	37.1	7.70	25.0	86	24.18	65.2	31.89	11.6
41	7.47	37.9	8.08	24.4	87	24.56	65.6	32.55	11.5
42	7.80	38.7	8.46	23.8	88	24.93	66.1	33.21	11.4
43	8.13	39.5	8.85	23.3	89	25.30	66.5	33.87	11.2
44	8.47	40.3	9.25	22.7	90	25.67	66.9	34.57	11.1
45	8.82	41.0	9.67	22.2	91	26.04	67.3	35.21	11.0

Table 2. Field table for compass and degree abney.

Degrees	100x Vers	100x Sine	100x Tan	100x Cotan	100x Exsec	100x Cosec	100x Cos
0	0.00	0.00	0.00	----	0.00	------	100.00
1	0.02	1.75	1.75	5728.99	0.02	5729.87	99.98
2	0.06	3.49	3.49	2863.63	0.06	2865.37	99.94
3	0.14	5.23	5.24	1908.11	0.14	1910.73	99.86
4	0.24	6.98	6.99	1403.07	0.24	1433.56	99.76
5	0.38	8.72	8.75	1143.01	0.38	1147.37	99.62
6	0.55	10.45	10.51	951.44	0.55	956.68	99.45
7	0.75	12.49	12.28	814.43	0.75	820.55	99.25
8	0.87	13.92	14.05	711.54	0.98	718.53	99.03
9	1.23	15.64	15.84	631.38	1.25	639.25	98.77
10	1.52	17.36	17.63	567.13	1.54	575.88	98.48
11	1.84	19.08	19.44	514.46	1.87	524.08	98.16
12	2.19	20.79	21.26	470.46	2.23	480.97	97.81
13	2.56	22.50	23.09	433.15	2.63	444.54	97.44
14	2.97	24.19	24.93	401.08	3.06	413.36	97.03
15	3.41	25.88	26.79	373.21	3.53	386.37	96.59
16	3.87	27.56	28.67	348.74	4.03	362.80	96.13
17	4.37	29.24	30.57	327.09	4.57	342.03	95.63
18	4.89	30.90	32.49	307.77	5.15	323.61	95.11
19	5.45	32.56	34.43	290.42	5.76	307.16	94.55
20	6.03	34.20	36.40	274.75	6.42	292.38	93.97
21	6.64	35.84	38.39	260.51	7.11	279.04	93.36
22	7.28	37.46	40.40	247.51	7.85	266.95	92.72
23	7.95	39.07	42.45	235.58	8.64	255.93	92.05
24	8.65	40.67	44.52	224.60	9.46	245.86	91.35
25	9.36	42.26	46.63	214.45	10.34	236.62	90.63
26	10.12	43.84	48.77	205.03	11.26	228.12	89.88
27	10.90	45.40	50.95	196.26	12.23	220.27	89.10
28	11.70	46.95	53.17	188.07	13.26	213.01	88.29
29	12.54	48.48	55.43	180.40	14.34	206.27	87.46
30	13.40	50.00	57.74	173.21	15.47	200.00	86.60
31	14.28	51.50	60.09	166.43	16.66	194.16	88.72
32	15.19	52.99	62.49	160.03	17.92	188.71	84.80
33	16.13	54.46	64.94	153.99	19.24	183.61	83.87
34	17.10	55.92	67.45	148.26	20.62	178.83	82.90
35	18.08	57.36	70.02	142.81	22.08	174.34	81.92
36	19.10	58.77	72.65	137.64	23.61	170.13	80.90
37	20.14	60.18	75.36	132.70	25.21	166.16	79.86
38	21.20	61.57	78.13	127.99	26.90	162.43	78.80
39	22.29	62.93	80.98	123.49	28.68	158.90	77.71
40	23.40	64.28	83.91	119.18	30.54	155.57	76.40
41	24.53	65.61	86.93	115.37	32.50	152.43	75.47
42	25.69	66.91	90.04	111.06	34.56	146.63	74.31
43	26.86	68.20	93.25	107.24	36.73	146.63	73.14
44	28.07	69.47	96.57	103.55	39.02	143.96	71.93
45	29.29	70.71	100.00	100.00	41.42	141.42	70.71

Table 3. Conversion of degrees to percent and topographic graduations.

Degree	%	Topog.	Degree	%	Topog.
30′	0.87	0.58	20	36.40	24.02
1 00′	1.74	1.15	30′	37.39	24.68
1 30′	2.62	1.73	21	38.39	25.33
2 00′	3.49	2.30	30′	39.39	26.00
2 30′	4.37	2.88	22	40.40	26.67
3 00′	5.24	3.46	30′	41.42	27.34
3 30′	6.12	4.04	23	42.45	28.02
4 00′	6.99	4.62	30′	43.48	28.70
4 30′	7.87	5.19	24	44.52	29.39
5 00′	8.75	5.77	30′	45.57	30.08
5 30′	9.63	6.36	25	46.63	30.78
6 00′	10.51	6.94	30′	47.70	31.48
6 30′	11.39	7.52	26	48.77	32.19
7 00′	12.28	8.10	30′	49.86	32.91
7 30′	13.16	8.69	27	50.95	33.63
8 00′	14.05	9.28	30′	52.06	34.36
8 30′	14.94	9.86	28	53.17	35.09
9 00′	15.84	10.45	30′	54.30	35.84
30′	16.73	11.04	29	55.43	36.58
10	17.63	11.64	30′	56.58	37.34
30′	18.53	12.23	30	57.73	38.11
11	19.44	12.83	30′	58.90	38.88
30′	20.34	13.43	31	60.09	39.66
12	21.26	14.03	30′	61.28	40.44
30′	22.17	14.63	32	62.49	41.24
13	23.09	15.24	30′	63.71	42.05
30′	24.01	15.85	33	64.94	42.86
14	24.93	16.46	30′	66.19	43.68
30′	25.86	17.07	34	67.45	44.52
15	26.80	17.68	30′	68.73	45.36
30′	27.73	18.30	35	70.02	46.21
16	28.67	18.93	30′	71.33	47.08
30′	29.62	19.55	36	72.65	47.95
17	30.57	20.18	30′	74.00	48.84
30′	31.53	20.81	37	75.35	49.73
18	32.49	21.44	30′	76.73	50.64
30′	33.46	22.08	38	78.13	51.57
19	34.43	22.73	30′	79.54	52.50
30′	35.41	23.38			

(Continued)

Table 3. (Continued).

Degree	%	Topog.	Degree	%	Topog.
39	80.98	53.45	58	160.03	105.62
30′	82.43	54.41	30′	163.19	107.70
40	83.91	55.38	59	166.43	109.84
30′	85.41	56.37	30′	169.77	112.05
41	86.93	57.37	60	173.20	114.32
30′	88.47	58.39	30′	176.75	116.65
42	90.04	59.43	61	180.41	119.07
30′	91.63	60.48	30′	184.18	121.56
43	93.25	61.55	62	188.07	124.13
30′	94.90	62.63	30′	192.10	126.28
44	96.57	63.74	63	196.26	129.53
30′	98.27	64.86	30′	200.57	132.38
45	100.00	66.00	64	205.03	135.32
30′	101.76	67.16	30′	209.65	138.37
46	103.55	68.34	65	214.45	141.54
30′	105.38	69.55	70	274.75	181.33
47	107.24	70.78	75	373.20	246.32
30′	109.13	72.03	80	567.13	374.30
48	111.06	73.30	85	1143.01	754.39
30′	113.03	74.60			
49	115.04	75.92			
30′	117.08	77.28			
50	119.18	78.65			
30′	121.31	80.06			
51	123.49	81.50			
30′	125.72	82.97			
52	127.99	84.48			
30′	130.32	86.01			
53	132.70	87.58			
30′	135.14	89.19			
54	137.64	90.84			
30′	140.20	92.53			
55	142.81	94.26			
30′	145.50	96.03			
56	148.26	97.85			
30′	151.08	99.72			
57	153.99	101.63			
30′	156.97	103.60			

Table 4. Conversion of topographic graduations to percent and degrees.

Topog.	%	Degree		Topog.	%	Degree	
1.0	1.52	0°	52′	24.0	36.36	19°	59′
1.5	2.27	1	18′	24.5	37.12	20	22′
2.0	3.03	1	44′	25.0	37.88	20	45′
2.5	3.79	2	10′	25.5	38.64	21	08′
3.0	4.55	2	36′	26.0	39.39	21	30′
3.5	5.30	3	02′	26.5	40.15	21	53′
4.0	6.06	3	28′	27.0	40.91	22	15′
4.5	6.82	3	54′	27.5	41.67	22	37′
5.0	7.58	4	20′	28.0	42.42	22	59′
5.5	8.33	4	46′	28.5	43.18	23	21′
6.0	9.09	5	12′	29.0	43.94	23	43′
6.5	9.85	5	38′	29.5	44.70	24	05′
7.0	10.61	6	03′	30.0	45.45	24	27′
7.5	11.36	6	29′	30.5	46.21	24	48′
8.0	12.12	6	55′	31.0	46.97	25	10′
8.5	12.88	7	20′	31.5	47.73	25	31′
9.0	13.64	7	46′	32.0	48.48	25	52′
9.5	14.39	8	11′	32.5	49.24	26	13′
10.0	15.15	8	37′	33.0	50.00	26	34′
10.5	15.91	9	02′	33.5	50.76	26	55′
11.0	16.67	9	28′	34.0	51.51	27	15′
11.5	17.42	9	53′	34.5	52.27	27	36′
12.0	18.18	10	18′	35.0	53.03	27	56′
12.5	18.94	10	43′	35.5	53.79	28	16′
13.0	19.70	11	08′	36.0	54.55	28	37′
13.5	20.45	11	34′	36.5	55.30	28	57′
14.0	21.21	11	59′	37.0	56.06	29	17′
14.5	21.97	12	23′	37.5	56.82	29	36′
15.0	22.73	12	48′	38.0	57.58	29	56′
15.5	23.48	13	13′	38.5	58.33	30	15′
16.0	24.24	13	38′	39.0	50.09	30	35′
16.5	25.00	14	02′	39.5	59.85	30	54′
17.0	25.76	14	27′	40.0	60.60	31	13′
17.5	26.52	14	51′	40.5	61.36	31	32′
18.0	27.27	15	15′	41.0	62.12	31	51′
18.5	28.03	15	40′	41.5	62.88	32	10′
19.0	28.79	16	04′	42.0	63.64	32	28′
19.5	29.55	16	28′	42.5	64.39	32	47′
20.0	30.30	16	51′	43.0	65.15	33	05′
20.5	31.06	17	15′	43.5	65.91	33	23′
21.0	31.82	17	39′	44.0	66.67	33	41′
21.5	32.58	18	02′	44.5	67.42	33	59′
22.0	33.33	18	26′	45.0	68.18	34	17′
22.5	34.09	18	50′	45.5	68.94	34	35′
23.0	34.85	19	13′	46.0	69.70	34.	53′
23.5	35.61	19	36′	46.5	70.45	35	10′

(Continued)

Table 4. (Continued).

Topog.	%	Degree		Topog.	%	Degree	
47.0	71.21	35°	27′	71.0	107.58	47°	05′
47.5	71.97	35	45′	71.5	108.33	47	17′
48.0	72.73	36	02′	72.0	109.09	47	29′
48.5	73.48	36	19′	72.5	109.85	47	41′
49.0	74.24	36	35′	73.0	110.61	47	53′
49.5	75.00	36	52′	73.5	111.36	48	05′
50.0	75.76	37	09′	74.0	112.12	48	16′
50.5	76.52	37	25′	74.5	112.88	48	28′
51.0	77.27	37	42′	75.0	113.64	48	39′
51.5	78.03	37	58′	75.5	114.39	48	50′
52.0	78.79	38	14′	76.0	115.15	49	02′
52.5	79.55	38	30′	76.5	115.91	49	13′
53.0	80.30	38	46′	77.0	116.67	49	24′
53.5	81.06	39	02′	77.5	117.42	49	35′
54.0	81.82	39	17′	78.0	118.18	49	46′
54.5	82.58	39	33′	78.5	118.94	49	57′
55.0	83.33	39	48′	79.0	119.70	50	07′
55.5	84.09	40	04′	79.5	120.45	50	18′
56.0	84.85	40	19′	80.0	121.21	50	29′
56.5	85.61	40	34′	80.5	121.97	50	39′
57.0	86.36	40	49′	81.0	122.73	50	50′
57.5	87.12	41	04′	81.5	123.48	51	00′
58.0	87.88	41	19′	82.0	124.24	51	10′
58.5	88.64	41	33′	81.5	125.00	51	20′
59.0	89.39	41	48′	83.0	125.76	51	31′
59.5	90.15	42	02′	83.5	126.52	51	41′
60.0	90.91	42	16′	84.0	127.27	51	51′
60.5	91.66	42	31′	84.5	128.03	52	00′
61.0	92.42	42	45′	85.0	128.79	52	10′
61.5	93.18	42	59′	85.5	129.55	52	20′
62.0	93.94	43	13′	86.0	130.30	52	30′
62.5	94.69	43	26′	86.5	131.06	52	39′
63.0	95.45	43	40′	87.0	131.82	52	49′
63.5	96.21	43	54′	87.5	132.58	52	59′
64.0	96.97	44	07′	88.0	133.33	53	08′
64.5	97.73	44	20′	88.5	134.09	53	17′
65.0	98.48	44	34′	89.0	134.85	53	26′
65.5	99.24	44	47′	89.5	135.61	53	36′
66.0	100.00	45	00′	90.0	136.35	53	45′
66.5	100.76	45	13′	91.0	137.88	54	03′
67.0	101.52	45	26′	92.0	139.39	54	20′
67.5	102.27	45	39′	93.0	140.91	54	38′
68.0	103.03	45	51′	94.0	142.42	54	55′
68.5	103.79	46	05′	95.0	143.94	55	13′
69.0	104.55	46	16′	96.0	145.45	55	29′
69.5	105.30	46	29′	97.0	146.97	55	46′
70.0	106.06	46	41′	98.0	148.48	56	02′
70.5	106.82	46	53′	99.0	150.00	56	19′
				100.0	151.52	56	35′

Table 5. Conversion of percent to topographic graduations and degrees.

%	Topog.	Degrees		%	Topog.	Degrees	
1	0.66	0°	34′	20.0	13.20	11°	19′
1.5	0.99	0	52′	20.5	13.53	11	35′
2.0	1.32	1	09′	21.0	13.86	11	52′
2.5	1.65	1	26′	21.5	14.19	12	08′
3.0	1.98	1	43′	22.0	14.52	12	24′
3.5	2.31	2	00′	22.5	14.85	12	41′
4.0	2.64	2	17′	23.0	15.18	12	57′
4.5	2.97	2	35′	23.5	15.51	13	13′
5.0	3.30	2	52	24.0	15.84	13	30′
5.5	3.63	3	09′	24.5	16.17	13	46′
6.0	3.96	3	26′	25.0	16.50	14	02′
6.5	4.29	3	43′	25.5	16.83	14	18′
7.0	4.62	4	00′	26.0	17.16	14	34′
7.5	4.95	4	17′	26.5	17.49	14	51′
8.0	5.28	4	34′	27.0	17.82	15	07′
8.5	5.61	4	52′	27.5	18.15	15	23′
9.0	5.94	5	09′	28.0	18.48	15	39′
9.5	6.37	5	26′	28.5	18.81	15	54′
10.0	6.60	5	43′	29.0	19.14	16	10′
10.5	6.93	6	00′	29.5	19.47	16	26′
11.0	7.26	6	17′	30.0	19.80	16	42′
11.5	7.59	6	34′	30.5	20.13	16	58′
12.0	7.92	6	51′	31.0	20.46	17	13′
12.5	8.25	7	07′	31.5	20.79	17	29′
13.0	8.58	7	24′	32.0	21.12	17	45′
13.5	8.91	7	41′	32.5	21.45	18	00′
14.0	9.24	7	58′	33.0	21.78	18	16′
14.5	9.57	8	15′	33.5	22.11	18	31′
15.0	9.90	8	32′	34.0	22.44	18	47′
15.5	10.23	8	49′	34.5	22.77	19	02′
16.0	10.56	9	05′	35.0	23.10	19	17′
16.5	10.89	9	22′	35.5	23.43	19	33′
17.0	11.22	9	39′	36.0	23.76	19	48′
17.5	11.55	9	56′	36.5	24.09	20	03′
18.0	11.88	10	12′	37.0	24.42	20	18′
18.5	12.21	10	29′	37.5	24.75	20	33′
19.0	12.54	10	45′	38.0	25.08	20	48′
19.5	12.87	11	02′	38.5	25.41	21	03′

(Continued)

Table 5. (Continued).

%	Topog.	Degrees		%	Topog.	Degrees	
39.0	25.74	21°	18′	58.0	38.28	30°	07′
39.5	26.07	21	33′	58.5	38.61	30	20′
40.0	26.40	21	48′	59.0	38.94	30	32′
40.5	26.73	22	03′	59.5	39.27	30	45′
41.0	27.06	22	18′	60.0	39.60	30	58′
41.5	27.39	22	32′	60.5	39.93	31	10′
42.0	27.72	22	47′	61.0	40.26	31	23′
42.5	28.05	23	02′	61.5	40.59	31	35′
43.0	28.38	23	16′	62.0	40.92	31	48′
43.5	28.71	23	31′	62.5	41.25	32	00′
44.0	29.04	23	45′	63.0	41.58	32	13′
44.5	29.37	23	59′	63.5	41.91	32	25′
45.0	29.70	24	14′	64.0	42.24	32	37′
45.5	30.03	24	28′	64.5	42.57	32	49′
46.0	30.36	24	42′	65.0	42.90	33	01′
46.5	30.69	24	56′	65.5	43.23	33	13′
47.0	31.02	25	10′	66.0	43.56	33	25′
47.5	31.35	25	24′	66.5	43.89	33	37′
48.0	31.68	25	38′	67.0	44.22	33	49′
48.5	32.01	25	52′	67.5	44.55	34	01′
49.0	32.34	26	06′	68.0	44.88	34	13′
49.5	32.67	26	20′	68.5	45.21	34	25′
50.0	33.00	26	34′	69.0	45.54	34	36′
50.5	33.33	26	48′	69.5	45.87	34	48′
51.0	33.66	27	01′	70.0	46.20	35	00′
51.5	33.99	27	15′	70.5	46.53	35	11′
52.0	34.32	27	28′	71.0	46.86	35	22′
52.5	34.65	27	42′	71.5	47.19	35	34′
53.0	34.98	27	55′	72.0	47.52	35	45′
53.5	35.31	28	09′	72.5	47.85	35	57′
54.0	35.64	28	22′	73.0	48.18	36	08′
54.5	35.97	28	35′	73.5	48.51	36	19′
55.0	36.30	28	49′	74.0	48.84	36	30′
55.5	36.63	29	02′	74.5	49.17	36	41′
56.0	36.96	29	15′	75.0	49.50	36	52′
56.5	37.29	29	28′	75.5	49.83	37	03′
57.0	37.62	29	41′	76.0	50.16	37	14′
57.5	37.95	29	54′	76.5	50.49	37	25′

(Continued)

Table 5. (Continued).

%	Topog.	Degrees		%	Topog.	Degrees	
77.0	50.82	37°	36′	88.0	58.08	41°	21′
77.5	51.15	37	47′	88.5	58.41	41	31′
78.0	51.48	37	57′	89.0	58.74	41	40′
78.5	51.81	38	08′	89.5	59.07	41	50′
79.0	52.14	38	19′	90.0	59.40	41	59′
79.5	52.47	38	29′	90.5	59.73	42	09′
80.0	52.80	38	40′	91.0	60.06	42	18′
80.5	53.13	38	50′	91.5	60.39	42	28′
81.0	53.46	39	00′	92.0	60.72	42	37′
81.5	53.99	39	11′	92.5	61.05	42	46′
82.0	54.12	39	21′	93.0	61.38	42	55′
82.5	54.45	39	31′	93.5	61.71	43	05′
83.0	54.78	39	42′	94.0	62.04	43	14′
83.5	55.11	39	52′	94.5	62.37	43	23′
84.0	55.44	40	02′	95.0	62.70	43	32′
84.5	55.77	40	12′	95.5	63.03	43	41′
85.0	56.10	40	22′	96.0	63.36	43	50′
85.5	56.43	40	32′	96.5	63.69	43	59′
86.0	56.76	40	42′	97.0	64.02	44	08′
86.5	57.09	40	52′	97.5	64.35	44	17′
87.0	57.42	41	01′	98.0	64.68	44	25′
87.5	57.75	41	11′	98.5	65.01	44	34′
				99.0	65.34	44	43′
				99.5	65.67	44	51′
				100.0	66.00	45	00′

Slope distance in links	Topographic															
	5	10	15	20	25	30	35	40	45	50	55	60	65	70	75	80
2	2.0	2.0	2.0	1.9	1.9	1.8	1.8	1.7	1.7	1.6	1.5	1.5	1.4	1.4	1.3	1.3
4	4.0	4.0	4.0	3.8	3.7	3.6	3.5	3.4	3.3	3.2	3.1	3.0	2.8	2.7	2.6	2.5
6	6.0	5.9	5.9	5.7	5.6	5.5	5.3	5.1	5.0	4.8	4.6	4.4	4.3	4.1	4.0	3.8
8	8.0	7.9	7.8	7.7	7.5	7.3	7.1	6.8	6.6	6.4	6.1	5.9	5.7	5.5	5.3	5.1
10	10.0	9.9	9.8	9.6	9.4	9.1	8.8	8.6	8.3	8.0	7.7	7.4	7.1	6.9	6.6	6.4
12	12.0	11.9	11.7	11.5	11.2	10.9	10.6	10.3	9.9	9.6	9.2	8.9	8.5	8.2	7.9	7.6
14	14.0	13.8	13.7	13.4	13.1	12.7	12.4	12.0	11.6	11.2	10.8	10.4	10.0	9.6	9.2	8.9
16	16.0	15.8	15.6	15.3	15.0	14.6	14.1	13.7	13.2	12.8	12.3	11.8	11.4	11.0	10.6	10.2
18	17.9	17.8	17.6	17.2	16.8	16.4	15.9	15.4	14.9	14.3	13.8	13.3	12.8	12.3	11.9	11.5
20	19.9	19.8	19.5	19.1	18.7	18.2	17.7	17.1	16.5	15.9	15.4	14.8	14.2	13.7	13.2	12.7
22	21.9	21.8	21.5	21.0	20.6	20.0	19.4	18.8	18.2	17.5	16.9	16.3	15.7	15.1	14.5	14.0
24	23.9	23.7	23.4	23.0	22.4	21.8	21.2	20.5	19.8	19.1	18.4	17.8	17.1	16.5	15.9	15.3
26	25.9	25.7	25.4	24.9	24.3	23.7	23.0	22.2	21.5	20.7	20.0	19.2	18.5	17.8	17.2	16.5
28	27.9	27.7	27.3	26.8	26.2	25.5	24.7	23.9	23.1	22.3	21.5	20.7	19.9	19.2	18.5	17.8
30	29.9	29.7	29.3	28.7	28.1	27.3	26.5	25.7	24.8	23.9	23.0	22.2	21.4	20.6	19.8	19.1
32	31.9	31.6	31.2	30.6	29.9	29.1	28.3	27.4	26.4	25.5	24.6	23.7	22.8	22.0	21.1	20.4
34	33.9	33.6	33.2	32.5	31.8	31.0	30.0	29.1	28.1	27.1	26.1	25.2	24.2	23.3	22.5	21.6
36	35.9	35.6	35.1	34.5	33.7	32.8	31.8	30.8	29.7	28.7	27.7	26.6	25.6	24.7	23.8	22.9
38	37.9	37.6	37.1	36.4	35.5	34.6	33.6	32.5	31.4	30.3	29.2	28.1	27.1	26.1	25.1	24.2
40	39.9	39.5	39.0	38.3	37.4	36.4	35.3	34.2	33.0	31.9	30.7	29.6	28.5	27.4	26.4	25.5
42	41.9	41.5	41.0	40.2	39.3	38.2	37.0	35.9	34.7	33.5	32.3	31.1	29.9	28.8	27.7	26.7
44	43.9	43.5	42.9	42.1	41.1	40.1	38.9	37.6	36.4	35.1	33.8	32.6	31.3	30.2	29.1	28.0
46	45.9	45.5	44.9	44.0	43.0	41.9	40.6	39.3	38.0	36.7	35.3	34.0	32.8	31.6	30.4	29.3
48	47.9	47.5	46.8	45.9	44.9	43.7	42.4	41.0	39.7	38.3	36.9	35.3	34.2	32.9	31.7	30.5
50	49.9	49.4	48.8	47.9	46.8	45.5	44.2	42.8	41.3	39.9	38.4	37.0	35.6	34.3	33.0	31.8
52	51.9	51.4	50.7	49.8	48.6	47.3	45.9	44.5	43.0	41.4	39.9	38.5	37.0	35.7	34.4	33.1
54	53.8	53.4	52.7	51.7	50.5	49.2	47.7	46.2	44.6	43.0	41.5	40.0	38.5	37.0	35.7	34.4
56	55.8	55.4	54.6	53.6	52.4	51.0	49.5	47.9	46.3	44.6	43.0	41.4	39.9	38.4	37.0	35.6
58	57.8	57.3	56.6	55.5	54.2	52.8	51.2	49.6	47.9	46.2	44.6	42.9	41.3	39.8	38.3	36.9
60	59.8	59.3	58.5	57.4	56.1	54.6	53.0	51.3	49.6	47.8	46.1	44.4	42.7	41.2	39.6	38.2
62	61.8	61.3	60.5	59.3	58.0	56.4	54.8	53.0	51.2	49.4	47.6	45.9	44.2	42.5	41.0	39.5
64	63.8	63.3	62.4	61.2	59.9	58.3	56.5	54.7	52.9	51.0	49.2	47.4	45.6	43.9	42.3	40.7
66	65.8	65.3	64.4	63.2	61.7	60.1	58.3	56.4	54.5	52.6	50.7	48.8	47.0	45.3	43.6	42.0
68	67.8	67.2	66.3	65.1	63.6	61.9	60.1	58.1	56.2	54.2	52.2	50.3	48.4	46.4	44.9	43.3
70	69.8	69.2	68.3	67.0	65.5	63.7	61.8	59.9	57.8	55.8	53.8	51.8	49.9	48.0	46.2	44.5
72	71.8	71.2	70.2	68.9	67.3	65.5	63.6	61.6	59.5	57.4	55.3	53.3	51.3	49.4	47.6	45.8
74	73.8	73.2	72.2	70.8	69.2	67.4	65.4	63.3	61.1	59.0	56.8	54.8	52.7	50.8	48.9	47.1
76	75.8	75.1	74.1	72.7	71.1	69.2	67.1	65.0	62.8	60.6	58.4	56.2	54.1	52.1	50.2	48.4
78	77.8	77.1	76.1	74.6	72.9	71.0	68.9	66.7	64.4	62.2	59.9	57.7	55.6	53.5	51.5	49.6
80	79.8	79.1	78.0	76.6	74.8	72.8	70.7	68.4	66.1	63.8	61.5	59.2	57.0	54.9	52.9	50.9
82	81.8	81.1	80.0	78.5	76.6	74.6	72.4	70.1	67.8	65.4	63.0	60.7	58.4	56.3	54.2	52.2
84	83.8	81.3	81.9	80.4	78.6	76.5	74.2	71.8	69.4	67.0	64.5	62.2	59.8	57.6	55.5	53.5
86	85.8	85.0	83.9	82.3	80.4	78.3	76.0	73.5	71.1	68.5	66.1	63.6	61.3	59.0	56.8	54.7
88	87.7	87.0	85.8	84.2	82.3	80.1	77.7	75.3	72.7	70.1	67.6	65.1	62.7	60.4	58.1	56.0
90	89.7	89.0	87.8	86.1	84.2	81.9	79.5	77.0	74.4	71.7	69.1	66.6	64.1	61.7	59.5	57.3
92	91.7	91.0	89.7	88.0	86.0	83.8	81.3	78.7	76.0	73.3	70.7	68.1	65.5	63.1	60.8	58.5
94	93.7	92.9	91.7	90.0	87.9	85.6	83.0	80.4	77.7	74.9	72.2	69.6	67.0	64.7	62.1	59.8
96	95.7	94.9	93.6	91.9	89.8	87.4	84.8	82.1	79.3	76.5	73.7	71.0	68.4	65.9	63.4	61.1
98	97.7	96.9	95.6	93.8	91.6	89.2	86.6	83.8	81.0	78.1	75.3	72.5	69.8	67.2	64.7	62.4
100	99.7	98.9	97.5	95.7	93.5	91.0	88.3	85.5	82.6	79.7	76.8	74.0	71.2	68.6	66.1	63.6

Table 6. Conversion of slope distance to horizontal distance for various readings on the topographic abney.

Table 7. Slope reduction table for the percent abney. (Compiled by Robert L. Wilson)

	2%(1°09')		3%(1°43')		4%(2°17')		5%(2°52')	
S.D.	H.D.	D.E.	H.D.	D.E.	H.D.	D.E.	H.D.	D.E.
40	39.99	.80	39.98	1.20	39.97	1.60	39.95	2.00
45	44.99	.90	44.98	1.35	44.96	1.80	44.94	2.25
50	49.99	1.00	49.98	1.50	49.96	2.00	49.94	2.50
55	54.99	1.10	54.98	1.65	54.96	2.20	54.93	2.75
60	59.99	1.20	59.97	1.80	59.95	2.40	59.93	3.00
65	64.99	1.30	64.97	1.95	64.95	2.60	64.92	3.25
70	69.99	1.40	69.97	2.10	69.94	2.80	69.91	3.50
75	74.99	1.50	74.97	2.25	74.94	3.00	74.91	3.75
80	79.98	1.60	79.96	2.40	79.94	3.20	79.90	4.00
85	84.98	1.70	84.96	2.55	84.93	3.40	84.90	4.24
90	89.98	1.80	89.96	2.70	89.93	3.60	89.89	4.49
95	94.98	1.90	94.96	2.85	94.92	3.80	94.88	4.74
100	99.98	2.00	99.96	3.00	99.92	4.00	99.88	4.99
105	104.98	2.10	104.95	3.15	104.92	4.20	104.87	5.24
110	109.98	2.20	109.95	3.30	109.91	4.40	109.87	5.49
115	114.98	2.30	114.95	3.45	114.91	4.60	114.86	5.74
120	119.98	2.40	119.95	3.60	119.90	4.80	119.85	5.99
125	124.98	2.50	124.94	3.75	124.90	5.00	124.84	6.24
130	129.97	2.60	129.94	3.90	129.90	5.20	129.84	6.49
135	134.97	2.70	134.94	4.05	134.89	5.40	134.83	6.74
140	139.97	2.80	139.94	4.20	139.89	5.60	139.83	6.99
145	144.97	2.90	144.93	4.35	144.88	5.80	144.82	7.24
150	149.97	3.00	149.93	4.50	149.88	6.00	149.81	7.49
155	154.97	3.10	154.93	4.65	154.88	6.20	154.81	7.74
160	159.97	3.20	159.93	4.80	159.87	6.39	159.80	7.99
165	164.97	3.30	164.93	4.95	164.87	6.59	164.80	8.24
170	169.97	3.40	169.92	5.10	169.86	6.79	169.79	8.49
175	174.97	3.50	174.92	5.25	174.86	6.99	174.78	8.74
180	179.96	3.60	179.92	5.40	179.86	7.19	179.78	8.99
185	184.96	3.70	184.92	5.55	184.85	7.39	184.77	9.24
190	189.96	3.80	189.91	5.70	189.85	7.59	189.76	9.49
195	194.96	3.90	194.91	5.85	194.84	7.79	194.76	9.74
200	199.96	4.00	199.91	6.00	199.84	7.99	199.75	9.99
205	204.96	4.10	204.91	6.15	204.84	8.19	204.74	10.24
210	209.96	4.20	209.91	6.30	209.83	8.39	209.74	10.49
215	214.96	4.30	214.90	6.45	214.83	8.59	214.73	10.74
220	219.96	4.40	219.90	6.60	219.82	8.79	219.73	10.99
225	224.96	4.50	224.90	6.75	224.82	8.99	224.72	11.24
230	229.96	4.60	229.90	6.90	229.82	9.19	229.71	11.49
235	234.95	4.70	234.89	7.05	234.81	9.39	234.71	11.74
240	239.95	4.80	239.89	7.20	239.81	9.59	239.70	11.99
245	244.95	4.90	244.89	7.35	244.80	9.79	244.70	12.23
250	249.95	5.00	249.89	7.50	249.80	9.99	249.69	12.48
255	254.95	5.10	254.89	7.65	254.80	10.19	254.69	12.73
260	259.95	5.20	259.88	7.80	259.79	10.39	259.68	12.98
265	264.95	5.30	264.88	7.95	264.79	10.59	264.67	13.23
270	269.95	5.40	269.88	8.10	269.78	10.79	269.66	13.48
275	269.95	5.50	274.88	8.25	274.78	10.99	274.66	13.73
280	279.94	5.60	279.87	8.40	279.78	11.19	279.65	13.98
285	284.94	5.70	284.87	8.55	284.77	11.39	284.64	14.23
290	289.94	5.80	289.87	8.70	289.77	11.59	289.64	14.48
295	294.94	5.90	294.87	8.85	294.76	11.79	294.63	14.73
300	299.94	6.00	299.87	9.00	299.76	11.99	299.63	14.98
350	349.93	7.00	349.84	10.50	349.72	13.99	349.56	17.48

S.D.	6% (3° 26′) H.D.	6% (3° 26′) D.E.	7% (4° 00′) H.D.	7% (4° 00′) D.E.	8% (4° 34′) H.D.	8% (4° 34′) D.E.	9% (5° 09′) H.D.	9% (5° 09′) D.E.
40	39.93	2.40	39.90	2.79	39.87	3.18	39.84	3.59
45	44.92	2.69	44.89	3.14	44.86	3.58	44.82	4.04
50	49.92	2.99	49.87	3.49	49.84	3.97	49.80	4.49
55	54.90	3.29	54.87	3.84	54.84	4.38	54.78	4.94
60	59.89	3.59	59.85	4.19	59.81	4.78	59.76	5.39
65	64.88	3.89	64.84	4.53	64.79	5.17	64.74	5.83
70	69.88	4.19	69.83	4.88	69.78	5.57	69.72	6.28
75	74.87	4.49	74.82	5.23	74.76	5.97	74.70	6.73
80	79.86	4.79	79.81	5.58	79.75	6.37	79.68	7.18
85	84.85	5.09	84.79	5.93	84.73	6.77	84.66	7.63
90	89.84	5.39	89.78	6.28	89.71	7.16	89.64	8.08
95	94.83	5.69	94.77	6.63	94.70	7.56	94.62	8.53
100	99.82	5.99	99.76	6.98	99.68	7.96	99.60	8.98
105	104.81	6.28	104.74	7.32	104.67	8.36	104.58	9.42
110	109.80	6.58	109.73	7.67	109.65	8.76	109.56	9.87
115	114.79	6.89	114.71	8.02	114.63	9.16	114.54	10.32
120	119.78	7.19	119.71	8.37	119.62	9.55	119.52	10.77
125	124.78	7.49	124.70	8.72	124.60	9.95	124.50	11.22
130	129.77	7.79	129.68	9.07	129.59	10.35	129.47	11.67
135	134.76	8.08	134.67	9.42	134.57	10.75	134.46	12.12
140	139.75	8.38	139.66	9.77	139.55	11.15	139.43	12.57
145	144.74	8.68	144.65	10.11	144.54	11.54	144.41	13.02
150	149.73	8.93	149.63	10.46	149.52	11.94	149.39	13.46
155	154.72	9.28	154.62	10.81	154.51	12.34	154.37	13.91
160	159.71	9.58	159.61	11.16	159.49	12.74	159.35	14.36
165	164.70	9.88	164.60	11.50	164.48	13.14	164.33	14.81
170	169.69	10.18	169.59	11.85	169.46	13.53	169.31	15.26
175	174.69	10.48	174.57	12.21	174.44	13.93	174.29	15.71
180	179.68	10.78	179.56	12.56	179.43	14.33	179.27	16.16
185	184.67	11.08	184.55	12.90	184.41	14.73	184.25	16.61
190	189.66	11.38	189.54	13.25	189.40	15.13	189.23	17.05
195	194.65	11.68	194.52	13.60	194.38	15.53	194.21	17.50
200	199.64	11.98	199.51	13.95	199.36	15.92	199.19	17.95
205	204.63	12.28	204.50	14.30	204.35	16.32	204.17	18.40
210	209.62	12.58	209.49	14.65	209.33	16.72	209.15	18.85
215	214.61	12.88	214.48	15.00	214.32	17.12	214.13	19.30
220	219.61	13.18	219.46	15.35	219.30	17.52	219.11	19.75
225	224.60	13.47	224.45	15.70	224.28	17.91	224.09	20.20
230	229.59	13.77	229.44	16.04	229.27	18.31	229.07	20.64
235	234.58	14.07	234.43	16.39	234.25	18.71	234.05	21.09
240	239.57	14.37	239.42	16.79	239.24	19.11	239.03	21.54
245	244.56	14.67	244.40	17.09	244.22	19.51	224.01	21.99
250	249.55	14.97	249.39	17.44	249.21	19.90	248.99	22.44
255	254.54	15.27	254.38	17.79	254.19	20.30	253.97	22.89
260	259.53	15.57	259.37	18.14	259.17	20.70	258.95	23.34
265	264.52	15.87	264.35	18.49	264.16	21.10	263.93	23.79
270	269.51	16.17	269.34	18.83	269.14	21.50	268.91	24.24
275	274.51	16.47	274.33	19.18	274.13	21.90	273.89	24.68
280	279.50	16.77	279.32	19.53	279.11	22.29	278.87	25.13
285	284.49	17.07	284.31	19.88	284.09	22.69	283.85	25.82
290	289.48	17.37	289.29	20.23	289.08	23.09	288.83	26.03
295	294.47	17.67	294.28	20.58	294.06	23.49	293.81	26.48
300	299.48	17.97	299.27	20.93	299.05	23.89	298.79	26.93
350	349.37	20.96	349.15	24.44	348.89	27.91	348.59	31.37

S.D.	10% (5° 43′) H.D.	D.E.	11% (6° 17′) H.D.	D.E.	12% (6° 51′) H.D.	D.E.	13% (7° 24′) H.D.	D.E.
40	39.80	3.98	39.76	4.38	39.71	4.77	39.67	5.15
45	44.78	4.48	44.73	4.92	44.68	5.37	44.63	5.80
50	49.75	4.98	49.70	5.47	49.64	5.96	49.58	6.44
55	54.73	5.48	54.67	6.02	54.61	6.56	54.54	7.08
60	59.70	5.98	59.64	6.56	59.57	7.16	59.50	7.73
65	64.68	6.47	64.61	7.11	64.54	7.75	64.46	8.37
70	69.65	6.97	69.58	7.66	69.50	8.35	69.42	9.02
75	74.63	7.47	74.55	8.20	74.46	8.95	74.38	9.66
80	79.60	7.97	79.52	8.75	79.43	9.54	79.33	10.30
85	84.58	8.47	84.49	9.30	84.39	10.14	84.29	10.95
90	89.55	8.96	89.46	9.85	89.36	10.73	89.25	11.59
95	94.53	9.46	94.43	10.39	94.32	11.33	94.21	12.24
100	99.50	9.96	99.40	10.94	99.29	11.93	99.17	12.88
105	104.48	10.46	104.37	11.49	104.25	12.52	104.13	13.52
110	109.45	10.96	109.34	12.03	109.21	13.12	109.08	14.17
115	114.43	11.45	114.31	12.58	114.18	13.72	114.04	14.81
120	119.40	11.95	119.28	13.13	119.14	14.31	119.00	15.46
125	124.38	12.45	124.25	13.67	124.11	14.91	123.96	16.10
130	129.35	12.95	129.22	14.22	129.07	15.51	128.92	16.74
135	134.33	13.45	134.19	14.77	134.04	16.10	133.88	17.39
140	139.30	13.94	139.16	15.32	139.00	16.70	138.83	18.03
145	144.28	14.44	144.13	15.86	143.96	17.29	143.79	18.68
150	149.25	14.94	149.10	16.41	148.93	17.89	148.75	19.32
155	154.23	15.44	154.07	16.96	153.89	18.49	153.71	19.96
160	159.20	15.94	159.04	17.50	158.86	19.08	158.67	20.61
165	164.18	16.43	164.01	18.05	163.82	19.68	163.63	21.25
170	169.15	16.93	168.98	18.60	168.79	20.28	168.58	21.90
175	174.13	17.43	173.95	19.14	173.75	20.87	173.54	22.54
180	179.10	17.93	178.92	19.69	178.72	21.47	178.50	23.18
185	184.08	18.43	183.89	20.24	183.68	22.06	183.46	23.83
190	189.05	18.92	188.86	20.79	188.64	22.66	188.42	24.47
195	194.03	19.42	193.83	21.33	193.61	23.26	193.38	25.12
200	199.00	19.92	198.80	21.88	198.57	23.85	198.33	25.76
205	203.98	20.42	203.77	22.43	203.54	24.45	203.29	26.40
210	208.95	20.92	208.74	22.97	208.50	25.05	208.25	27.05
215	213.93	21.41	213.71	23.52	213.47	25.64	213.21	27.69
220	218.90	21.91	218.68	24.07	218.43	26.24	218.17	28.34
225	223.88	22.41	223.65	24.61	223.39	26.83	223.13	28.98
230	228.85	22.91	228.62	25.16	228.36	27.43	228.08	29.62
235	233.83	23.41	233.59	25.71	233.32	28.03	233.04	30.27
240	238.80	23.90	238.56	26.26	238.29	28.62	238.00	30.91
245	243.78	24.40	243.53	26.80	243.25	29.22	242.96	31.55
250	248.75	24.90	248.50	27.35	248.22	29.82	247.92	32.20
255	253.73	25.40	253.47	27.90	253.18	30.41	252.88	32.84
260	258.71	25.90	258.44	28.44	258.14	31.01	257.83	33.49
265	263.68	26.39	263.41	28.99	263.11	31.61	262.79	34.13
270	268.66	26.89	268.38	29.54	268.07	32.20	267.75	34.77
275	273.62	27.39	273.35	30.08	273.04	32.80	272.71	35.42
280	278.61	27.89	278.32	30.63	278.00	33.40	277.67	36.06
285	283.58	28.39	283.29	31.18	282.97	33.99	282.63	36.71
290	288.56	28.88	288.26	31.73	287.93	34.59	287.58	37.35
295	293.53	29.38	293.23	32.27	292.89	35.18	292.54	37.99
300	298.51	29.88	298.20	32.82	297.86	35.78	297.50	38.64
350	348.26	34.83	347.90	38.27	347.51	41.70	347.08	45.12

	14% (7° 58')		15% (8° 32')		16% (9° 05')		17% (9° 39')	
S.D.	H.D.	D.E.	H.D.	D.E.	H.D.	D.E.	H.D.	D.E.
40	39.61	5.54	39.56	5.94	39.50	6.31	39.42	6.71
45	44.57	6.24	44.50	6.68	44.44	7.10	44.36	7.54
50	49.52	6.93	49.45	7.42	49.37	7.89	49.29	8.38
55	54.47	7.62	54.39	8.16	54.31	8.68	54.22	9.22
60	59.42	8.32	59.34	8.90	59.25	9.47	59.15	10.06
65	64.37	9.01	64.28	9.65	64.18	10.26	64.08	10.90
70	69.32	9.70	69.23	10.39	69.12	11.05	69.01	11.73
75	74.28	10.39	74.17	11.13	74.06	11.84	73.94	12.57
80	79.23	11.09	79.11	11.87	79.00	12.63	78.88	13.41
85	84.18	11.78	84.06	12.61	83.93	13.42	83.80	14.25
90	89.13	12.47	89.00	13.35	88.87	14.21	88.73	15.09
95	94.08	13.17	93.95	14.10	93.81	15.00	93.66	15.92
100	99.03	13.86	98.89	14.84	98.75	15.79	98.59	16.76
105	103.99	14.55	103.84	15.58	103.68	16.58	103.51	17.60
110	108.94	15.25	108.78	16.32	108.62	17.37	108.44	18.44
115	113.89	15.94	113.73	17.06	113.56	18.16	113.37	19.28
120	118.84	16.63	118.67	17.81	118.49	18.94	118.30	20.12
125	123.79	17.32	123.62	18.55	123.43	19.73	123.23	20.95
130	128.75	18.02	128.56	19.29	128.37	20.52	128.16	21.79
135	133.70	18.71	133.51	20.03	133.31	21.31	133.09	22.63
140	138.65	19.40	138.45	20.77	138.24	22.10	138.02	23.47
145	143.60	20.10	143.39	21.52	143.18	22.89	142.95	24.31
150	148.55	20.79	148.34	22.26	148.12	23.68	147.88	25.14
155	153.50	21.48	153.28	23.00	153.06	24.47	152.81	25.98
160	158.46	22.18	158.23	23.74	157.99	25.25	157.74	26.82
165	163.41	22.87	163.17	24.48	162.93	26.05	162.67	27.66
170	168.36	23.56	168.12	25.23	167.87	26.84	167.59	28.50
175	173.31	24.25	173.06	25.97	172.80	27.63	172.52	29.33
180	178.26	24.95	178.01	26.71	177.74	28.42	177.45	30.17
185	183.21	25.64	182.95	27.45	182.68	29.21	182.38	31.01
190	188.17	26.33	187.90	28.19	187.62	30.00	187.31	31.85
195	193.12	27.03	192.84	28.94	192.55	30.78	192.24	32.69
200	198.07	27.72	197.79	29.68	197.49	31.57	197.17	33.53
205	203.02	28.41	202.73	30.42	202.43	31.57	202.10	32.36
210	207.97	29.11	207.68	31.16	207.36	33.15	207.03	35.20
215	212.93	29.80	212.62	31.90	212.30	33.94	211.96	36.04
220	217.88	30.49	217.56	32.64	217.24	34.73	216.89	36.88
225	222.83	31.18	222.51	33.39	222.18	35.52	221.82	37.72
230	227.78	31.88	227.45	34.13	227.11	36.31	226.75	38.55
235	232.73	32.57	232.40	34.87	232.05	37.10	231.67	39.39
240	237.68	33.26	237.34	35.61	236.99	37.89	236.60	40.23
245	242.64	33.96	242.29	36.35	241.93	38.68	241.53	41.07
250	247.59	34.65	247.23	37.10	246.86	39.47	246.46	41.91
255	252.54	35.34	252.18	37.84	251.80	40.26	251.39	42.75
260	257.49	36.04	257.12	38.58	256.74	41.05	256.32	43.58
265	262.44	36.73	262.07	39.32	261.68	41.84	261.25	44.42
270	267.39	37.42	267.01	40.06	266.61	42.63	266.18	45.26
275	272.35	38.11	271.96	40.81	271.55	43.41	271.11	46.10
280	277.30	38.81	276.90	41.55	276.49	44.20	276.04	46.94
285	282.25	39.50	281.84	42.30	281.43	44.99	280.97	47.77
290	287.20	40.19	286.79	43.03	286.36	45.78	285.90	48.61
295	292.15	40.89	291.73	43.77	291.30	46.57	290.83	49.45
300	297.10	41.60	296.68	44.52	296.24	47.36	295.76	50.29
350	346.62	48.53	346.13	51.92	345.60	55.30	345.05	58.66

	18% (10° 12')		19% (10° 45')		20% (11° 19')		21% (11° 52')	
S.D.	H.D.	D.E.	H.D.	D.E.	H.D.	D.E.	H.D.	D.E.
40	39.37	7.08	39.30	7.46	39.22	7.85	39.15	8.23
45	44.29	7.97	44.21	8.39	44.13	8.83	44.04	9.25
50	49.21	8.85	49.12	9.33	49.00	9.81	48.93	10.28
55	54.13	9.74	54.03	10.26	53.93	10.79	53.82	11.31
60	59.05	10.63	58.95	11.19	58.83	11.77	58.72	12.33
65	63.97	11.51	63.86	12.12	63.74	12.76	63.61	13.37
70	68.89	12.40	68.77	13.06	68.64	13.74	68.50	14.39
75	73.81	13.28	73.68	13.99	73.54	14.72	73.40	15.42
80	78.74	14.17	78.60	14.92	78.44	15.70	78.29	16.45
85	83.66	15.05	83.51	15.85	83.35	16.68	83.18	17.48
90	88.58	15.94	88.42	16.79	88.25	17.66	88.08	18.51
95	93.50	16.82	93.33	17.72	93.15	18.64	92.97	19.54
100	98.42	17.71	98.25	18.65	98.06	19.62	97.86	20.56
105	103.34	18.59	103.16	19.59	102.96	20.60	102.76	21.59
110	108.26	19.48	108.07	20.52	107.86	21.59	107.65	22.62
115	113.18	20.36	112.98	21.45	112.76	22.57	112.54	23.65
120	118.10	21.25	117.89	22.38	117.67	23.55	117.44	24.68
125	123.02	22.14	122.81	23.32	122.57	24.53	122.33	25.70
130	127.95	23.02	127.72	24.25	127.47	25.51	127.22	26.73
135	132.87	23.91	132.63	25.18	132.38	26.49	132.11	27.76
140	137.79	24.79	137.54	26.11	137.28	27.47	137.01	28.79
145	142.71	25.68	142.46	27.05	142.18	28.45	141.90	29.82
150	147.63	26.56	147.37	27.98	147.08	29.43	146.79	30.85
155	152.55	27.45	152.28	28.91	151.99	30.42	151.69	31.87
160	157.47	28.33	157.19	29.84	156.89	31.40	156.58	32.90
165	162.39	29.22	162.10	30.78	161.79	32.38	161.47	33.93
170	167.31	30.10	167.02	31.71	166.69	33.36	166.37	34.96
175	172.23	31.00	171.93	32.64	171.60	34.34	171.26	35.99
180	177.16	31.88	176.84	33.57	176.50	35.32	176.15	37.01
185	182.08	32.76	181.75	34.51	181.40	36.30	181.05	38.04
190	187.00	33.65	186.67	35.44	186.31	37.28	185.94	39.07
195	191.92	34.53	191.58	36.37	191.21	38.27	190.83	40.10
200	196.84	35.42	196.49	37.30	196.11	39.25	195.73	41.13
205	201.76	36.30	201.40	38.24	201.01	40.23	200.62	42.16
210	206.68	37.19	206.31	39.17	205.92	41.21	205.51	43.18
215	211.60	38.07	211.23	40.10	210.82	42.19	210.41	44.21
220	216.52	38.96	216.14	41.04	215.72	43.17	215.30	45.24
225	221.44	39.84	221.05	41.97	220.63	44.15	220.19	46.27
230	226.36	40.73	225.96	42.90	225.53	45.13	225.08	47.30
235	231.29	41.61	230.88	43.83	230.43	46.11	229.98	48.32
240	236.21	42.50	235.79	44.77	235.33	47.10	234.87	49.35
245	241.13	43.39	240.70	45.70	240.24	48.08	239.76	50.38
250	246.05	44.27	245.61	46.63	245.14	49.06	244.66	51.41
255	250.97	45.16	250.52	47.56	250.04	50.04	249.55	52.44
260	255.89	46.04	255.44	48.50	254.94	51.02	254.44	53.47
265	260.81	46.93	260.35	49.43	259.85	52.00	259.34	54.49
270	265.73	47.81	265.26	50.36	264.75	52.98	264.23	55.52
275	270.65	48.70	270.17	51.29	269.65	53.96	269.12	56.55
280	275.57	49.58	275.09	52.23	274.56	54.94	274.02	57.58
285	280.50	50.47	280.00	53.16	279.46	55.93	278.91	58.61
290	285.42	51.35	284.91	54.09	284.36	56.91	283.80	59.63
295	290.34	52.24	289.82	55.02	289.26	57.89	288.70	60.66
300	295.26	53.13	294.74	55.96	294.17	58.87	293.59	61.69
350	344.46	62.00	343.85	65.33	343.20	68.64	342.53	71.93

	22% (12° 24′)		23% (12° 57′)		24% (13° 30′)		25% (14° 02′)	
S.D.	H.D.	D.E.	H.D.	D.E.	H.D.	D.E.	H.D.	D.E.
40	39.07	8.59	38.98	8.96	38.89	9.34	38.81	9.69
45	43.95	9.66	43.86	10.08	43.76	10.50	43.66	10.90
50	48.83	10.74	48.73	11.21	48.62	11.67	48.51	12.11
55	53.72	11.81	53.60	12.32	53.48	12.84	53.36	13.32
60	58.60	12.88	58.47	13.45	58.34	14.00	58.21	14.53
65	63.48	13.96	63.35	14.57	63.20	15.17	63.06	15.74
70	68.37	15.03	68.22	15.69	68.07	16.34	67.92	16.95
75	73.25	16.11	73.09	16.81	72.93	17.50	72.77	18.17
80	78.13	17.18	77.97	17.93	77.79	18.67	77.62	19.38
85	83.02	18.25	82.84	19.05	82.65	18.84	82.47	20.59
90	87.90	19.33	87.71	20.17	87.51	21.01	87.32	21.80
95	92.78	20.40	92.58	21.28	92.38	22.17	92.17	23.01
100	97.67	21.47	97.46	22.41	97.24	23.34	97.02	24.22
105	102.55	22.55	102.33	23.53	102.10	24.51	101.87	25.43
110	107.43	23.62	107.20	24.65	106.96	25.67	106.72	26.64
115	112.32	24.69	112.08	25.77	111.82	26.84	111.58	27.85
120	117.20	25.77	116.95	26.89	116.68	28.01	116.43	29.06
125	122.08	26.84	121.82	28.01	121.55	29.17	121.28	30.28
130	126.97	27.92	126.69	29.13	126.41	30.34	126.13	31.49
135	131.85	28.99	131.57	30.25	131.27	31.51	130.98	32.70
140	136.73	30.06	136.44	31.37	136.13	32.68	135.83	33.91
145	141.62	31.14	141.31	32.49	140.99	33.84	140.68	35.12
150	146.50	32.21	146.18	33.62	145.86	35.01	145.53	36.33
155	151.38	33.28	151.06	34.74	150.72	36.17	150.38	37.54
160	156.27	34.36	155.93	35.86	155.58	37.34	155.24	38.75
165	161.15	35.43	160.80	36.98	160.44	38.51	160.09	39.96
170	166.03	36.51	165.68	38.10	165.30	39.68	164.94	41.17
175	170.92	37.58	170.55	39.22	170.16	40.84	169.79	42.39
180	175.80	38.65	175.42	40.34	175.03	42.01	174.64	43.60
185	180.68	39.73	180.29	41.46	179.89	43.18	179.49	44.81
190	185.57	40.80	185.17	42.58	184.75	44.35	184.34	46.02
195	190.45	41.87	190.04	43.70	189.61	45.51	189.19	47.23
200	195.33	42.95	194.91	44.82	194.47	46.68	194.05	48.44
205	200.22	44.02	199.79	45.94	199.34	47.85	198.90	49.65
210	205.10	45.09	204.66	47.06	204.20	49.01	203.75	50.86
215	209.98	46.17	209.53	48.18	209.06	50.18	208.60	52.07
220	214.87	47.24	214.40	49.30	213.92	51.35	213.45	53.28
225	219.75	48.32	219.28	50.42	218.78	52.51	218.30	54.50
230	224.63	49.39	224.15	51.54	223.65	53.68	213.45	55.71
235	229.52	50.46	229.02	52.66	228.51	54.85	228.00	56.92
240	234.40	51.54	233.90	53.78	233.37	56.02	232.85	58.13
245	239.28	52.61	238.77	54.90	238.23	57.18	237.71	59.34
250	244.17	53.68	243.64	56.03	243.09	58.35	242.56	60.55
255	249.05	54.76	248.51	57.15	247.95	59.52	247.41	61.76
260	253.93	55.83	253.39	58.27	252.82	60.68	252.26	62.97
265	258.82	56.90	258.26	59.39	257.68	61.85	257.11	64.18
270	263.70	57.98	263.13	60.51	262.54	63.02	261.96	65.40
275	268.58	59.05	268.01	61.63	267.40	64.18	266.81	66.61
280	273.47	60.13	272.88	62.75	272.26	65.35	271.66	67.82
285	278.35	61.20	277.75	63.87	277.13	66.52	276.51	69.03
290	283.23	62.27	282.62	64.99	281.99	67.69	281.37	70.24
295	288.12	63.35	287.50	66.11	286.85	68.85	286.22	71.45
300	293.00	64.42	292.37	67.23	291.71	70.02	291.07	72.66
350	341.83	75.20	341.09	78.45	340.34	81.68	339.56	84.89

S.D.	26% (14° 34') H.D.	D.E.	27% (15° 07') H.D.	D.E.	28% (15° 39') H.D.	D.E.	29% (16° 10') H.D.	D.E.
40	38.71	10.06	38.62	10.43	38.52	10.79	38.42	11.14
45	43.55	11.32	43.44	11.74	43.33	12.14	43.22	12.53
50	48.39	12.58	48.27	13.04	48.15	13.49	48.02	13.92
55	53.23	13.83	53.10	14.34	52.96	14.84	52.83	15.31
60	58.07	15.09	57.92	15.65	57.78	16.19	57.63	16.71
65	62.91	16.35	62.75	16.95	62.59	17.53	62.43	18.10
70	67.75	17.61	67.58	18.25	67.40	18.88	67.23	19.49
75	72.59	18.86	72.40	19.56	72.22	20.23	72.03	20.88
80	77.43	20.12	77.20	20.86	77.03	21.58	76.84	22.27
85	82.27	21.38	82.06	22.17	81.85	22.93	81.64	23.67
90	87.11	22.64	86.88	23.47	86.66	24.28	86.44	25.06
95	91.95	23.89	91.71	24.77	91.48	25.63	91.24	26.45
100	96.79	25.15	96.54	26.08	96.29	26.98	96.05	27.84
105	101.62	26.41	101.37	27.38	101.11	28.32	100.85	29.24
110	106.46	27.67	106.19	28.69	105.92	29.67	105.65	30.63
115	111.30	28.92	111.02	29.99	110.74	31.02	110.45	32.02
120	116.14	30.18	115.85	31.29	115.55	32.37	115.25	33.41
125	120.98	31.44	120.67	32.60	120.37	33.72	120.06	34.80
130	125.82	32.70	125.50	33.90	125.18	35.07	124.86	36.20
135	130.66	33.95	130.33	35.21	130.00	36.42	129.66	37.59
140	135.50	35.21	135.16	36.51	134.81	37.77	134.46	38.98
145	140.34	36.47	139.98	37.81	139.62	39.12	139.27	40.37
150	145.18	37.73	144.81	39.12	144.44	40.46	144.07	41.76
155	150.02	38.98	149.65	40.42	149.25	41.81	148.87	43.16
160	154.86	40.24	154.46	41.73	154.07	43.16	153.67	44.55
165	159.70	41.50	159.29	43.03	158.88	44.51	158.48	45.94
170	164.54	42.76	164.12	44.33	163.70	45.86	163.28	47.33
175	169.37	44.01	168.94	45.64	168.51	47.21	168.08	48.73
180	174.21	45.27	173.77	46.94	173.33	48.56	172.88	50.12
185	179.05	46.53	178.60	48.25	178.14	49.91	177.68	51.51
190	183.89	47.79	183.43	49.55	182.96	51.25	182.49	52.90
195	188.73	49.04	188.25	50.85	187.77	52.60	187.29	54.29
200	193.57	50.30	193.08	52.16	192.59	53.95	192.09	55.69
205	198.41	51.56	197.91	53.46	197.40	55.30	196.89	57.08
210	203.25	52.82	202.73	54.76	202.21	56.65	201.70	58.47
215	208.09	54.07	207.56	56.07	207.03	58.00	206.50	59.86
220	212.93	55.33	212.39	57.37	211.84	59.35	211.30	61.26
225	217.77	56.59	217.21	58.68	216.66	60.70	216.10	62.65
230	222.61	57.85	222.04	59.98	221.47	62.04	220.90	64.04
235	227.45	59.10	226.87	61.28	226.29	63.39	225.71	65.43
240	232.28	60.36	231.70	62.59	231.10	64.74	230.51	66.82
245	237.12	61.62	236.52	63.89	235.92	66.09	235.31	68.22
250	241.96	62.88	241.35	65.20	240.73	67.44	240.11	69.61
255	246.80	64.13	246.18	66.50	245.55	68.79	244.92	71.00
260	251.64	65.39	251.00	67.80	250.36	70.14	249.72	72.39
265	256.48	66.65	255.83	69.11	255.18	71.49	254.52	73.78
270	261.32	67.91	260.66	70.41	259.99	72.84	259.32	75.18
275	266.16	69.16	265.48	71.72	264.81	74.18	264.13	76.57
280	271.00	70.42	270.31	73.02	269.62	75.53	268.93	77.96
285	275.84	71.68	275.14	74.32	274.43	76.88	273.73	79.35
290	280.68	72.94	279.97	75.63	279.25	78.23	278.53	80.75
295	285.52	74.19	284.79	76.93	284.06	79.58	283.33	82.14
300	290.36	75.45	289.62	78.24	288.88	80.93	288.14	83.53
350	338.74	88.07	337.90	91.23	337.04	94.37	336.15	97.48

S.D.	30% (16° 42') H.D.	D.E.	31% (17° 13') H.D.	D.E.	32% (17° 45') H.D.	D.E.	33% (18° 16') H.D.	D.E.
40	38.31	11.49	38.21	11.84	38.10	12.19	37.98	12.54
45	43.10	12.93	42.98	13.32	42.86	13.72	42.73	14.10
50	47.89	14.37	47.80	14.80	47.62	15.24	47.48	15.67
55	52.68	15.80	52.54	16.28	52.38	16.77	52.23	17.24
60	57.47	17.24	57.31	17.76	57.12	18.29	56.98	18.81
65	62.26	18.68	62.09	19.24	61.91	19.82	61.72	20.37
70	67.05	20.12	66.86	20.72	66.67	21.34	66.47	21.94
75	71.84	21.55	71.64	22.20	71.43	22.86	71.22	23.51
80	76.63	22.99	76.42	23.69	76.19	24.39	75.97	25.07
85	81.41	24.43	81.19	25.16	80.95	25.91	80.72	26.64
90	86.20	25.86	85.97	26.64	85.72	27.44	85.46	28.21
95	90.99	27.30	90.74	28.12	90.48	28.96	90.21	29.78
100	95.78	28.74	95.52	29.60	95.24	30.49	94.96	31.34
105	100.57	30.17	100.30	31.08	100.00	32.01	99.71	32.91
110	105.36	31.61	105.07	32.56	104.76	33.54	104.46	34.48
115	110.15	33.05	109.85	34.04	109.53	35.06	109.20	36.05
120	114.94	34.48	114.62	35.52	114.29	36.58	114.95	37.61
125	119.73	35.92	119.40	37.00	119.05	38.11	118.70	39.18
130	124.52	37.36	124.17	38.48	123.81	39.63	123.45	40.75
135	129.31	38.79	128.95	39.96	128.57	41.16	128.20	42.31
140	134.09	40.23	133.73	41.44	133.34	42.68	132.95	43.88
145	138.88	41.67	138.50	42.92	138.10	44.21	137.69	45.45
150	143.67	43.10	143.28	44.40	142.86	45.73	142.44	47.02
155	148.46	44.54	148.05	45.88	147.62	47.25	147.19	48.58
160	153.25	45.98	152.83	47.36	152.38	48.78	151.94	50.15
165	158.04	47.41	157.61	48.84	157.15	50.30	156.69	51.72
170	162.83	48.85	162.38	50.32	161.91	51.83	161.43	53.28
175	167.62	50.29	167.16	51.80	166.67	53.35	166.18	54.85
180	172.41	51.72	171.93	53.28	171.43	54.87	170.93	56.42
185	177.20	53.16	176.71	54.76	176.19	56.40	175.68	57.99
190	181.99	54.60	181.49	56.24	180.96	57.92	180.43	59.55
195	186.78	56.04	186.26	57.72	185.72	59.45	185.17	61.12
200	191.56	57.47	191.04	59.20	190.48	60.97	189.92	62.69
205	196.35	58.91	195.81	60.68	195.24	62.50	194.67	64.26
210	201.14	60.35	200.59	62.16	200.00	64.02	199.42	65.82
215	205.93	61.78	205.37	63.64	204.77	65.55	204.17	67.40
220	210.72	63.22	210.14	65.12	209.53	67.07	208.91	68.96
225	215.51	64.66	214.92	66.60	214.29	68.59	213.66	70.52
230	220.30	66.09	219.69	68.08	219.05	70.12	218.41	72.09
235	225.09	67.53	224.47	69.56	223.81	71.64	223.16	73.66
240	229.88	68.97	229.25	71.04	228.57	73.17	227.91	75.23
245	234.67	70.40	234.02	72.52	233.34	74.69	232.65	76.79
250	239.46	71.84	238.80	74.00	238.10	76.22	237.40	78.36
255	244.24	73.28	243.57	75.48	242.86	77.74	242.15	79.93
260	249.03	74.71	248.35	76.96	247.62	79.26	246.90	81.49
265	253.82	76.15	253.13	78.44	252.38	80.79	251.65	83.06
270	258.61	77.59	257.90	79.92	257.15	82.31	256.39	84.63
275	263.40	79.02	262.68	81.40	261.91	83.84	261.14	86.20
280	268.19	80.46	267.45	82.88	266.67	85.40	265.89	87.76
285	272.98	81.90	272.23	84.36	271.42	86.89	270.64	89.33
290	277.77	83.33	277.06	85.84	276.19	88.41	275.39	90.90
295	282.56	84.77	281.78	87.32	280.96	89.93	280.13	92.46
300	287.35	86.21	286.56	88.80	285.72	91.46	284.88	94.03
350	335.24	100.57	334.31	103.63	333.35	106.67	332.37	109.68

	34% (18° 47′)		35% (19° 17′)		36% (19° 48′)		37% (20° 18′)	
S.D.	H.D.	D.E.	H.D.	D.E.	H.D.	D.E.	H.D.	D.E.
40	37.87	12.88	37.76	13.21	37.64	13.55	37.51	13.88
45	42.60	14.49	42.48	14.86	42.34	15.24	42.21	15.61
50	47.34	16.10	47.19	16.51	47.04	16.94	46.89	17.35
55	52.07	17.71	51.91	18.16	51.75	18.63	51.58	19.08
60	56.80	19.32	56.63	19.81	56.45	20.32	56.27	20.82
65	61.54	20.93	61.35	21.47	61.16	22.01	60.96	22.55
70	66.27	22.54	66.07	23.12	65.86	23.71	65.65	24.29
75	71.01	24.15	70.79	24.78	70.57	25.40	70.34	26.02
80	75.74	25.76	75.51	26.42	75.27	27.10	75.03	27.75
85	80.47	27.37	80.23	28.07	79.97	28.79	79.72	29.49
90	85.21	28.98	84.95	29.72	84.68	30.49	84.41	31.22
95	89.94	30.59	89.67	31.37	89.38	32.18	89.10	32.96
100	94.67	32.20	94.39	33.02	94.09	33.87	93.78	34.69
105	99.41	33.81	99.11	34.68	98.79	35.56	98.48	36.43
110	104.14	35.42	103.83	36.33	103.50	37.26	103.17	38.20
115	108.88	37.02	108.55	37.98	108.20	38.95	107.86	39.90
120	113.61	38.64	113.27	39.63	112.91	40.65	112.55	41.63
125	118.34	40.25	117.99	41.28	117.61	42.34	117.24	43.37
130	123.07	41.86	122.71	42.93	122.31	44.04	121.93	45.10
135	127.81	43.47	127.43	44.58	127.02	45.72	126.62	46.84
140	132.54	45.08	132.15	46.23	131.72	47.42	131.30	48.57
145	137.28	46.69	136.87	47.88	136.43	49.11	135.99	50.31
150	142.01	48.30	141.58	49.54	141.13	50.81	140.68	52.04
155	146.75	49.91	146.30	51.19	145.84	52.50	145.37	53.78
160	151.47	51.52	151.02	52.84	150.54	54.20	150.06	55.51
165	156.21	53.13	155.74	54.49	155.25	55.89	154.75	57.24
170	160.95	54.74	160.46	56.14	159.95	57.59	159.44	58.97
175	165.68	56.35	165.18	57.79	164.65	59.27	164.13	60.71
180	170.41	57.96	169.90	59.44	169.36	60.97	168.82	62.45
185	175.15	59.57	174.62	61.09	174.06	62.66	173.51	64.18
190	179.88	61.18	179.34	62.75	178.77	64.36	178.20	65.92
195	184.61	62.79	184.06	64.40	183.47	66.05	182.89	67.65
200	189.35	64.40	188.78	66.05	188.18	67.75	187.58	69.38
205	194.08	66.01	193.50	67.70	192.88	69.43	192.27	71.12
210	198.82	67.62	198.22	69.35	197.58	71.13	196.96	72.86
215	203.55	69.23	202.94	71.00	202.29	72.82	201.65	74.59
220	208.28	70.84	207.66	72.65	206.99	74.52	206.34	76.33
225	213.02	72.45	212.38	74.30	211.70	76.21	211.03	78.06
230	217.75	74.06	217.10	75.96	216.40	77.91	215.71	79.79
235	222.48	75.67	221.82	77.61	221.11	79.59	220.40	81.53
240	227.22	77.28	226.54	79.26	225.81	81.30	225.09	83.26
245	231.95	78.89	231.25	80.91	230.52	82.98	229.78	85.00
250	236.68	80.50	235.97	82.56	235.22	84.68	234.47	86.73
255	241.42	82.11	240.69	84.21	239.92	86.37	239.16	88.47
260	246.15	83.72	245.41	85.86	244.63	88.07	243.85	90.20
265	250.89	85.33	250.13	87.51	249.33	89.75	248.54	91.94
270	255.62	86.94	254.85	89.16	254.04	91.46	253.23	93.67
275	260.35	88.55	259.57	90.82	258.74	93.14	257.92	95.41
280	265.09	90.16	264.29	92.47	263.45	94.85	262.61	97.14
285	269.82	91.77	269.01	94.12	268.15	96.53	267.30	98.88
290	274.56	93.38	273.73	95.77	272.86	98.22	271.99	100.61
295	279.29	94.99	278.45	97.42	277.56	99.92	276.68	102.35
300	284.02	96.60	283.17	99.07	282.26	101.61	281.37	104.08
350	331.37	112.67	330.35	115.62	329.31	118.55	328.25	121.45

S.D.	38% (20° 48′) H.D.	38% (20° 48′) D.E.	39% (21° 18′) H.D.	39% (21° 18′) D.E.	40% (21° 48′) H.D.	40% (21° 48′) D.E.	41% (22° 18′) H.D.	41% (22° 18′) D.E.
40	37.39	14.20	37.27	14.53	37.14	14.85	37.01	15.18
45	42.07	15.98	41.93	16.35	41.78	16.71	41.63	17.08
50	46.74	17.76	46.58	18.16	46.42	18.57	46.26	18.97
55	51.42	19.53	51.24	19.98	51.07	20.43	50.89	20.87
60	56.09	21.31	55.90	21.80	55.71	22.28	55.61	22.77
65	60.76	23.08	60.56	23.61	60.35	24.14	60.14	24.66
70	65.44	24.86	65.22	25.43	64.99	26.00	64.76	26.56
75	70.11	26.63	69.88	27.24	69.64	27.85	69.39	28.46
80	74.79	28.41	74.54	29.06	74.28	29.71	74.02	30.36
85	79.46	30.18	79.19	30.88	78.92	31.57	78.64	32.25
90	84.13	31.96	83.85	32.69	83.56	33.42	83.27	34.15
95	88.81	33.74	88.51	34.51	88.21	35.28	87.89	36.05
100	93.48	35.51	93.17	36.33	92.85	37.14	92.52	37.95
105	98.16	37.29	97.83	38.14	97.49	38.99	97.15	39.84
110	102.83	39.06	102.49	39.96	102.13	40.85	101.77	41.74
115	107.50	40.84	107.14	41.77	106.78	42.71	106.40	43.64
120	112.18	42.61	111.80	43.59	111.42	44.56	111.03	45.53
125	116.85	44.39	116.46	45.41	116.06	46.42	115.65	47.43
130	121.53	46.16	121.12	47.22	120.70	48.28	120.28	49.33
135	126.20	47.94	125.78	49.04	125.35	50.13	124.90	51.23
140	130.88	49.71	130.44	50.85	129.99	51.99	129.53	53.12
145	135.55	51.49	135.10	52.67	134.63	53.85	134.16	55.02
150	140.22	53.27	139.75	54.49	139.27	55.71	138.78	56.92
155	144.90	55.04	144.41	56.30	143.92	57.56	143.41	58.82
160	149.57	56.82	149.07	58.12	148.56	59.42	148.03	60.71
165	154.25	58.59	153.73	59.94	153.20	61.28	152.66	62.61
170	158.92	60.37	158.39	61.75	157.84	63.13	157.28	64.51
175	163.59	62.14	163.05	63.57	162.49	64.99	161.91	66.40
180	168.27	63.92	167.70	65.39	167.13	66.85	166 54	68.30
185	172.94	65.69	172.36	67.20	171.77	68.70	171.16	70.20
190	177.62	67.47	177.02	69.02	176.41	70.56	175.79	72.10
195	182.29	69.25	181.68	70.83	181.05	72.42	180.42	73.99
200	186.97	71.02	186.34	72.65	185.70	74.27	185.04	75.89
205	191.64	72.80	191.00	74.47	190.34	76.13	189.67	77.79
210	196.31	74.57	195.66	76.28	194.98	77.99	194.29	79.68
215	200.99	76.35	200.31	78.10	199.62	79.84	198.92	81.58
220	205.66	78.12	204.97	79.92	204.27	81.70	203.55	83.48
225	210.34	79.90	209.63	81.73	208.91	83.56	208.17	85.38
230	215.01	81.67	214.29	83.55	213.55	85.41	212.80	87.27
235	219.68	83.45	218.95	85.36	218.19	87.27	217.42	89.17
240	224.36	85.23	223.61	87.18	222.84	89.13	222.05	91.07
245	229.03	87.00	228.26	89.00	227.48	90.99	226.68	92.97
250	233.71	88.78	232.92	90.81	232.12	92.84	231.30	94.86
255	238.38	90.55	237.58	92.63	236.76	94.70	235.93	96.76
260	243.05	92.33	242.24	94.44	241.41	96.56	240.55	98.66
265	247.73	94.10	246.90	96.26	246.05	98.41	245.18	100.56
270	252.40	95.88	251.56	98.08	250.69	100.27	249.81	102.45
275	257.08	97.65	256.22	99.89	255.33	102.13	254.43	104.35
280	261.75	99.43	260.87	101.71	259.98	103.98	259.06	106.25
285	266.43	101.21	265.53	103.53	264.62	105.84	263.68	108.15
290	271.10	102.98	270.19	105.34	269.26	107.70	268.31	110.04
295	275.77	104.76	274.85	107.16	273.90	109.55	272.94	111.94
300	280.45	106.53	279.51	108.98	278.55	111.41	277.56	113.84
350	327.17	124.33	326.08	127.17	324.97	129.99	323.84	132.77

S.D.	42% (22° 47′) H.D.	42% (22° 47′) D.E.	43% (23° 16′) H.D.	43% (23° 16′) D.E.	44% (23° 45′) H.D.	44% (23° 45′) D.E.	45% (24° 14′) H.D.	45% (24° 14′) D.E.
40	36.88	15.49	36.75	15.80	36.61	16.11	36.48	16.42
45	41.49	17.43	41.34	17.78	41.19	18.12	41.03	18.47
50	46.10	19.36	45.93	19.75	45.77	20.14	45.60	20.52
55	50.71	21.30	50.53	21.73	50.34	22.15	50.15	22.57
60	55.32	23.23	55.12	23.70	54.92	24.16	54.71	24.63
65	59.93	25.17	59.71	25.68	59.50	26.18	59.27	26.68
70	64.54	27.11	64.31	27.65	64.07	28.19	63.83	28.73
75	69.15	29.04	68.90	29.63	68.65	30.21	68.39	30.78
80	73.76	30.98	73.49	31.60	73.22	32.22	72.95	32.84
85	78.37	32.92	78.09	33.58	77.80	34.23	77.51	34.89
90	82.98	34.85	82.68	35.55	82.38	36.25	82.07	36.94
95	87.59	36.79	87.27	37.53	86.95	38.26	86.63	38.99
100	92.20	38.72	91.87	39.50	91.53	40.27	91.19	41.05
105	96.81	40.66	96.46	41.48	96.11	42.29	95.75	43.10
110	101.42	42.60	101.05	43.45	100.68	44.30	100.30	45.15
115	106.03	44.53	105.65	45.43	105.26	46.32	104.87	47.20
120	110.64	46.47	110.24	47.40	109.84	48.33	109.43	49.25
125	115.25	48.41	114.83	49.38	114.41	50.34	113.99	51.31
130	119.86	50.34	119.43	51.35	118.99	52.36	118.54	53.36
135	124.47	52.28	124.02	53.33	123.57	54.37	123.10	55.41
140	129.08	54.21	128.61	55.30	128.14	56.38	127.66	57.46
145	133.69	56.15	133.21	57.28	132.72	58.40	132.22	59.52
150	138.30	58.09	137.80	59.25	137.30	60.41	136.78	61.57
155	142.91	60.02	142.39	61.23	141.87	62.43	141.34	63.62
160	147.52	61.96	146.99	63.20	146.45	64.44	145.90	65.67
165	152.13	63.90	151.58	65.18	151.03	66.45	150.46	67.72
170	156.74	65.83	156.17	67.15	155.60	68.47	155.02	69.78
175	161.35	67.77	160.77	69.13	160.18	70.48	159.58	71.83
180	165.96	69.70	165.36	71.10	164.76	72.49	164.14	73.88
185	170.57	71.64	169.96	73.08	169.33	74.51	168.70	75.93
190	175.18	73.58	174.55	75.05	173.91	76.52	173.26	78.00
195	179.79	75.51	179.14	77.03	178.49	78.54	177.82	80.04
200	184.40	77.45	183.74	79.00	183.06	80.55	182.38	82.09
205	189.01	79.39	188.33	80.98	187.64	82.56	186.94	84.14
210	193.61	81.32	192.92	82.95	192.22	84.58	191.50	86.20
215	198.22	83.26	197.52	84.93	196.79	86.59	196.05	88.25
220	202.83	85.19	202.11	86.90	201.37	88.60	200.61	90.30
225	207.44	87.13	206.70	88.88	205.95	90.62	205.17	92.35
230	212.05	89.07	211.30	90.85	210.52	92.63	209.73	94.40
235	216.66	91.00	215.89	92.83	215.10	94.65	214.29	96.46
240	221.27	92.94	220.48	94.80	219.67	96.66	218.85	98.51
245	225.88	94.88	225.08	96.78	224.25	98.67	223.41	100.56
250	230.49	96.81	229.67	98.75	228.83	100.69	227.97	102.61
255	235.10	98.75	234.26	100.73	233.40	102.70	232.53	104.67
260	239.71	100.68	238.86	102.70	237.98	104.70	237.09	106.72
265	244.32	102.62	243.45	104.68	242.56	106.73	241.65	108.77
270	248.93	104.56	248.04	106.70	247.13	108.74	246.21	110.82
275	253.54	106.49	252.64	108.63	251.71	110.76	250.77	112.87
280	258.15	108.43	257.23	110.60	256.29	112.77	255.33	114.93
285	262.76	110.37	261.82	112.58	260.86	114.78	259.89	116.98
290	267.37	112.30	266.42	114.55	265.44	116.80	264.45	119.03
295	271.98	114.24	271.01	116.53	270.02	118.81	269.01	121.08
300	276.59	116.17	275.60	118.50	274.59	120.82	273.56	123.13
350	322.69	135.53	321.53	138.26	320.36	140.96	319.17	143.63

	46% (24° 42')		47% (25° 10')		48% (25° 38')		49% (26° 06')	
S.D.	H.D.	D.E.	H.D.	D.E.	H.D	D.E.	H.D.	D.E.
40	36.34	16.71	36.20	17.01	36.06	17.30	35.92	17.60
45	40.88	18.80	40.73	19.14	40.57	19.47	40.41	19.80
50	45.43	20.89	45.25	21.26	45.08	21.63	44.90	22.00
55	49.97	22.98	49.78	23.39	49.59	23.79	49.39	24.20
60	54.51	25.07	54.30	25.52	54.09	25.97	53.88	26.40
65	59.05	27.16	58.83	27.64	58.60	28.12	58.37	28.60
70	63.60	29.25	63.36	29.77	63.11	30.28	62.86	30.80
75	68.14	31.34	67.88	31.89	67.62	32.45	67.35	33.00
80	72.68	33.43	72.41	34.02	72.13	34.61	71.84	35.20
85	77.22	35.52	76.93	36.15	76.63	36.77	76.33	37.40
90	81.77	37.61	81.46	38.27	81.14	38.93	80.82	39.59
95	86.31	39.70	85.98	40.40	85.65	41.10	85.31	41.79
100	90.85	41.79	90.51	42.53	90.16	43.26	89.80	43.99
105	95.39	43.88	95.03	44.65	94.67	45.42	94.29	46.19
110	99.94	45.97	99.56	46.78	99.17	47.59	98.78	48.39
115	104.48	48.05	104.08	48.90	103.68	49.75	103.27	50.59
120	109.02	50.14	108.61	51.03	108.19	51.91	107.76	52.79
125	113.56	52.23	113.13	53.16	112.70	54.08	112.25	54.99
130	118.11	54.32	117.66	55.28	117.21	56.24	116.74	57.19
135	122.65	56.41	122.19	57.41	121.71	58.40	121.23	59.40
140	127.19	58.50	126.71	59.54	126.22	60.57	125.72	61.59
145	131.73	60.59	131.24	61.66	130.73	62.73	130.21	63.79
150	136.28	62.68	135.76	63.79	135.24	64.89	134.70	65.99
155	140.82	64.77	140.29	65.91	139.75	67.05	139.19	68.19
160	145.36	66.86	144.81	68.04	144.25	69.22	143.68	70.39
165	149.90	68.95	149.34	70.17	148.76	71.38	148.17	72.59
170	154.45	71.04	153.86	72.29	153.27	73.54	152.66	74.79
175	158.99	73.13	158.39	74.42	157.78	75.71	157.15	76.99
180	163.53	75.22	162.91	76.55	162.28	77.87	161.64	79.19
185	168.07	77.31	167.44	78.67	166.79	80.03	166.14	81.39
190	172.62	79.39	171.96	80.80	171.30	82.20	170.63	83.59
195	177.16	81.48	176.49	82.92	175.81	84.36	175.12	85.79
200	181.70	83.57	181.01	85.05	180.32	86.52	179.61	87.99
205	186.24	85.66	185.54	87.18	184.82	88.69	184.10	90.19
210	190.79	87.75	190.07	89.30	189.33	90.85	188.59	92.39
215	195.33	89.84	194.59	91.43	193.84	93.01	193.08	94.59
220	199.87	91.93	199.12	93.56	198.35	95.17	197.57	96.79
225	204.41	94.02	203.64	95.68	202.86	97.34	202.06	98.99
230	208.96	96.11	208.17	97.81	207.36	99.50	206.55	101.19
235	213.50	98.20	212.69	99.93	211.87	101.66	211.04	103.39
240	218.04	100.29	217.22	102.06	216.38	103.83	215.53	105.59
245	222.58	102.38	221.74	104.19	220.89	105.99	220.02	107.79
250	227.13	104.47	226.27	106.31	225.40	108.15	224.51	109.98
255	231.67	106.56	230.79	108.44	229.90	110.32	229.00	112.18
260	236.21	108.65	235.32	110.57	234.41	112.48	233.49	114.38
265	240.75	110.73	239.84	112.69	238.92	114.64	237.98	116.58
270	245.30	112.82	244.37	114.82	243.43	116.80	242.47	118.78
275	249.84	114.91	248.90	116.94	247.93	118.97	246.96	120.98
280	254.38	117.00	253.42	119.07	252.44	121.13	251.45	123.18
285	258.92	119.09	257.95	121.20	256.95	123.29	255.94	125.38
290	263.47	121.18	262.47	123.32	261.46	125.46	260.43	127.58
295	268.01	123.27	267.00	125.45	265.97	127.62	264.92	129.78
300	272.55	125.36	271.52	127.58	270.47	129.78	269.41	131.98
350	317.97	146.27	316.76	148.88	315.53	151.46	314.30	154.01

	50% (26° 34′)		51% (27° 1′)		52% (27° 28′)		53% (27° 55′)	
S.D.	H.D.	D.E.	H.D.	D.E.	H.D.	D.E.	H.D.	D.E.
40	35.78	17.89	35.63	18.17	35.49	18.45	35.35	18.73
45	40.25	20.13	40.09	20.44	39.93	20.76	39.76	21.07
50	44.72	22.36	44.54	22.71	44.36	23.06	44.18	23.41
55	49.19	24.60	49.00	24.98	48.80	25.37	48.60	25.75
60	53.66	26.83	53.45	27.25	53.24	27.67	53.02	28.09
65	58.14	29.07	57.91	29.53	57.67	29.98	57.44	30.43
70	62.61	31.31	62.36	31.80	62.11	32.29	61.85	32.77
75	67.08	33.54	66.82	34.07	66.55	34.59	66.27	35.11
80	71.55	35.78	71.27	36.34	70.98	36.90	70.69	37.45
85	76.03	38.02	75.72	38.61	75.42	39.20	75.11	39.80
90	80.50	40.25	80.18	40.88	79.86	41.51	79.53	42.14
95	84.97	42.49	84.63	43.15	84.29	43.82	83.94	44.48
100	89.44	44.72	89.09	45.42	88.73	46.12	88.36	46.82
105	93.91	46.96	93.54	47.70	93.16	48.43	92.78	49.16
110	98.39	49.20	98.00	49.97	97.60	50.74	97.20	51.50
115	102.86	51.43	102.45	52.24	102.04	53.04	101.62	53.84
120	107.33	53.67	106.90	54.51	106.47	55.35	106.04	56.18
125	111.80	55.90	111.36	56.78	110.91	57.65	110.45	58.52
130	116.27	58.14	115.81	59.05	115.35	59.96	114.87	60.86
135	120.75	60.38	120.27	61.32	119.78	62.27	119.29	63.21
140	125.22	62.61	124.72	63.59	124.22	64.57	123.71	65.55
145	129.69	64.85	129.18	65.87	128.66	66.88	128.13	67.89
150	134.16	67.09	133.63	68.14	133.09	69.18	132.54	70.23
155	138.63	69.32	138.09	70.41	137.53	71.49	136.96	72.57
160	143.11	71.56	142.54	72.68	141.96	73.80	141.38	74.91
165	147.58	73.79	146.99	74.95	146.40	76.10	145.80	77.25
170	152.05	76.03	151.45	77.22	150.84	78.41	150.22	79.59
175	156.52	78.27	155.90	79.49	155.27	80.72	154.64	81.93
180	160.99	80.50	160.36	81.76	159.71	83.02	159.05	84.27
185	165.47	82.74	164.81	84.04	164.15	85.33	163.47	86.61
190	169.94	84.97	169.27	86.31	168.58	87.63	167.89	88.96
195	174.41	87.21	173.72	88.58	173.02	89.94	172.31	91.30
200	178.88	89.45	178.17	90.85	177.46	92.25	176.73	93.64
205	183.35	91.68	182.63	93.12	181.89	94.55	181.14	95.98
210	187.83	93.92	187.08	95.39	186.33	96.85	185.56	98.32
215	192.30	96.16	191.54	97.66	190.77	99.16	189.98	100.66
220	196.77	98.39	195.99	99.93	195.20	101.47	194.40	103.00
225	201.24	100.63	200.45	102.21	199.64	103.78	198.82	105.34
230	205.72	102.86	204.90	104.48	204.07	106.08	203.23	107.68
235	210.19	105.10	209.36	106.75	208.51	108.39	207.65	110.02
240	214.66	107.34	213.81	109.02	212.95	110.70	212.07	112.36
245	219.13	109.57	218.26	111.29	217.38	113.00	216.49	114.71
250	223.60	111.81	222.72	113.56	221.82	115.31	220.91	117.05
255	228.08	114.05	227.17	115.83	226.26	117.61	225.33	119.39
260	232.55	116.28	231.63	118.10	230.69	119.92	229.74	121.73
265	237.02	118.52	236.08	120.38	235.13	122.23	234.16	124.07
270	241.49	120.75	240.54	122.65	239.57	124.53	238.58	126.41
275	245.96	122.99	244.99	124.92	244.00	126.84	243.00	128.75
280	250.44	125.23	249.44	127.19	248.44	129.15	247.42	131.09
285	254.91	127.46	253.90	129.46	252.87	131.45	251.83	133.43
290	259.38	129.70	258.35	131.73	257.31	133.76	256.25	135.77
295	263.85	131.94	262.81	134.00	261.75	136.06	260.67	138.12
300	268.32	134.17	267.26	136.27	266.18	138.37	265.09	140.46
350	313.05	156.52	311.79	159.01	310.53	161.47	309.25	163.90

	54% (28° 22')		55% (28° 49')		56% (29° 15')		57% (29° 41')	
S.D.	H.D.	D.E.	H.D.	D.E.	H.D.	D.E.	H.D.	D.E.
40	35.20	19.00	35.05	19.28	34.90	19.54	34.75	19.81
45	39.60	21.38	39.43	21.69	39.26	21.99	39.09	22.28
50	44.00	23.76	43.81	24.10	43.62	24.43	43.44	24.76
55	48.40	26.13	48.19	26.51	47.99	26.87	47.78	27.24
60	52.80	28.51	52.57	28.92	52.35	29.32	52.13	28.71
65	57.20	30.88	56.95	31.33	56.71	31.76	56.47	32.19
70	61.59	33.26	61.33	33.74	61.07	34.20	60.81	34.66
75	65.99	35.63	65.71	36.15	65.44	36.65	65.16	37.14
80	70.39	38.01	70.09	38.56	69.80	39.09	69.50	39.62
85	74.79	40.38	74.47	40.97	74.16	41.53	73.85	42.09
90	79.19	42.76	78.85	43.38	78.52	43.98	78.19	44.57
95	83.59	45.14	83.24	45.79	82.89	46.42	82.53	47.04
100	87.99	47.51	87.62	48.20	87.25	48.86	86.88	49.52
105	92.39	49.89	92.00	50.61	91.61	51.31	91.22	52.00
110	96.79	52.26	96.38	53.02	95.97	53.75	95.57	54.47
115	101.19	54.64	100.76	55.43	100.34	56.19	99.91	56.95
120	105.59	57.01	105.14	57.84	104.70	58.63	104.25	59.42
125	109.99	59.39	109.52	60.25	109.06	61.08	108.60	61.90
130	114.39	61.76	113.90	62.66	113.42	63.52	112.94	64.38
135	118.79	64.14	118.28	65.07	117.79	65.96	117.28	66.85
140	123.19	66.52	122.66	67.48	122.15	68.41	121.63	69.33
145	127.59	68.89	127.04	69.89	126.51	70.85	125.97	71.80
150	131.99	71.27	131.42	72.30	130.87	73.29	130.32	74.28
155	136.39	73.64	135.81	74.71	135.24	75.74	134.66	76.76
160	140.79	76.02	140.19	77.12	139.60	78.18	139.00	79.23
165	145.19	78.39	144.57	79.53	143.96	80.62	143.35	81.71
170	149.59	80.77	148.95	81.94	148.32	83.07	147.69	84.19
175	153.99	83.14	153.33	84.35	152.69	85.51	152.04	86.66
180	158.39	85.52	157.71	86.76	157.05	87.95	156.38	89.14
185	162.79	87.90	162.09	89.17	161.41	90.39	160.72	91.61
190	167.19	90.27	166.47	91.58	165.77	92.84	165.07	94.09
195	171.59	92.65	170.85	93.99	170.14	95.28	169.41	96.57
200	175.99	95.02	175.23	96.40	174.50	97.72	173.76	99.04
205	180.38	97.40	179.61	98.81	178.86	100.17	178.10	101.52
210	184.78	99.77	183.99	101.22	183.22	102.61	182.44	103.99
215	189.18	102.15	188.38	103.63	187.59	105.05	186.79	106.47
220	193.58	104.52	192.76	106.04	191.95	107.50	191.13	108.95
225	197.98	106.90	197.14	108.45	196.31	109.94	195.47	111.42
230	202.38	109.28	201.52	110.86	200.67	112.38	199.82	113.90
235	206.78	111.65	205.90	113.27	205.04	114.83	204.16	116.37
240	211.18	114.03	210.28	115.68	209.40	117.27	208.51	118.85
245	215.58	116.40	214.66	118.09	213.76	119.71	212.85	121.33
250	219.98	118.79	219.04	120.50	218.12	122.16	217.19	123.80
255	224.38	121.15	223.42	122.91	222.49	124.60	221.54	126.28
260	228.78	123.53	227.80	125.32	226.85	127.04	225.88	128.75
265	233.18	125.90	232.18	127.73	231.21	129.48	230.23	131.23
270	237.58	128.28	236.56	130.14	235.57	131.93	234.57	133.71
275	241.98	130.66	240.95	132.55	239.94	134.37	238.91	136.18
280	246.38	133.03	245.33	134.96	244.30	136.81	243.26	138.66
285	250.78	135.41	249.71	137.37	248.66	139.26	247.60	141.13
290	255.18	137.78	254.09	139.78	253.02	141.70	251.94	143.61
295	259.58	140.16	258.47	142.19	257.39	144.14	256.29	146.09
300	263.98	142.53	262.85	144.60	261.75	146.59	260.63	148.56
350	307.97	166.30	306.67	168.67	305.38	171.01	304.07	173.32

S.D.	58% (30° 07') H.D.	D.E.	59% (30° 32') H.D.	D.E.	60% (30° 58') H.D.	D.E.	61% (31° 23') H.D.	D.E.
40	34.60	20.07	34.45	20.32	34.30	20.58	34.15	20.83
45	38.93	22.58	38.76	22.86	38.59	23.15	38.42	23.43
50	43.25	25.09	43.07	25.40	42.87	25.72	42.69	26.04
55	47.58	27.60	47.37	27.94	47.16	28.30	46.95	28.64
60	51.90	30.11	51.68	30.48	51.45	30.87	51.22	31.25
65	56.23	32.61	55.99	33.02	55.74	33.45	55.49	33.85
70	60.55	35.12	60.29	35.56	60.02	36.02	59.76	36.45
75	64.88	37.63	64.60	38.10	64.31	38.59	64.03	39.06
80	69.20	40.14	68.91	40.64	68.60	41.16	68.30	41.66
85	73.53	42.65	73.21	43.18	72.88	43.74	72.56	44.26
90	77.85	45.16	77.52	45.72	77.17	46.31	76.83	46.87
95	82.18	47.67	81.83	48.26	81.46	48.88	81.10	49.47
100	86.50	50.18	86.13	50.80	85.75	51.45	85.37	52.08
105	90.83	52.69	90.44	53.34	90.03	54.03	89.64	54.68
110	95.15	55.19	94.75	55.88	94.32	56.60	93.91	57.28
115	99.48	57.70	99.05	58.42	98.61	59.17	98.18	59.89
120	103.80	60.21	103.36	60.96	102.90	61.74	102.44	62.49
125	108.13	62.72	107.67	63.50	107.18	64.32	106.71	65.10
130	112.45	65.23	111.97	66.05	111.47	66.89	110.98	67.70
135	116.78	67.74	116.28	68.59	115.76	69.46	115.25	70.30
140	121.10	70.25	120.59	71.13	120.05	72.04	119.52	72.91
145	125.43	72.76	124.89	73.67	124.33	74.61	123.79	75.51
150	129.75	75.26	129.20	76.21	128.62	77.18	128.06	78.11
155	134.08	77.77	133.51	78.75	132.91	79.75	132.32	80.72
160	138.40	80.28	137.81	81.29	137.19	82.33	136.59	83.32
165	142.73	82.79	142.12	83.83	141.48	84.90	140.86	85.93
170	147.05	85.30	146.43	86.37	145.77	87.47	145.13	88.53
175	151.38	87.81	150.73	88.91	150.06	90.04	149.40	91.13
180	155.70	90.32	155.04	91.45	154.34	92.62	153.67	93.74
185	160.03	92.83	159.35	93.99	158.63	95.19	157.93	96.34
190	164.35	95.33	163.65	96.53	162.92	97.76	162.20	98.94
195	168.68	97.84	167.96	99.07	167.21	100.34	166.47	101.55
200	173.00	100.35	172.27	101.61	171.49	102.91	170.74	104.15
205	177.33	102.86	176.57	104.15	175.78	105.48	175.01	106.76
210	181.65	105.37	180.88	106.69	180.07	108.05	179.28	109.36
215	185.98	107.88	185.19	109.23	184.36	110.63	183.55	111.96
220	190.30	110.39	189.49	111.77	188.64	113.20	187.81	114.57
225	194.63	112.90	193.80	114.31	192.93	115.77	192.08	117.17
230	198.95	115.41	198.11	116.85	197.22	118.34	196.35	119.78
235	203.28	117.91	202.41	119.39	201.50	120.92	200.62	122.39
240	207.60	120.42	206.72	121.93	205.79	123.50	204.89	124.98
245	211.93	122.93	211.03	124.47	210.08	126.06	209.16	127.59
250	216.25	125.44	215.33	127.01	214.37	128.63	213.43	130.19
255	220.58	127.95	219.64	129.55	218.65	131.21	217.69	132.79
260	224.90	130.46	223.95	132.09	222.94	133.78	221.96	135.40
265	229.23	132.97	228.25	134.63	227.23	136.35	226.23	138.00
270	233.55	135.48	232.56	137.17	231.52	138.93	230.50	140.61
275	237.88	137.98	236.87	139.71	235.80	141.50	234.77	143.21
280	242.20	140.49	241.17	143.25	240.09	144.07	239.04	145.81
285	246.53	143.00	245.48	144.79	244.38	146.64	243.31	148.42
290	250.85	145.51	249.79	147.33	248.67	149.22	247.57	151.02
295	255.18	148.02	254.09	149.87	252.95	151.79	251.84	153.62
300	259.50	150.53	258.40	152.41	257.24	154.36	256.11	156.23
350	302.76	175.60	301.44	177.85	300.12	180.07	298.80	182.27

	62% (31° 48')		63% (32° 13')		64% (32° 37')		65% (33° 01')	
S.D.	H.D.	D.E.	H.D.	D.E.	H.D.	D.E.	H.D.	D.E.
40	34.00	21.08	33.84	21.32	33.69	21.56	33.54	21.80
45	38.25	23.71	38.07	23.99	37.90	24.26	37.73	24.52
50	42.49	26.35	42.30	26.65	42.11	26.95	41.93	27.24
55	46.74	28.98	46.53	29.32	46.33	29.65	46.12	29.97
60	50.99	31.62	50.76	31.99	50.54	32.34	50.31	32.69
65	55.24	34.25	54.99	34.65	54.75	35.04	54.50	35.42
70	59.49	36.89	59.22	37.32	58.96	37.73	58.70	38.14
75	63.74	39.52	63.45	39.98	63.17	40.43	62.89	40.87
80	67.99	42.16	67.68	42.65	67.38	43.12	67.08	43.59
85	72.24	44.79	71.91	45.31	71.60	45.82	71.27	46.32
90	76.49	47.43	76.14	47.98	75.81	48.51	75.47	49.04
95	80.74	50.06	80.37	50.64	80.02	51.21	79.66	51.76
100	84.99	52.70	84.60	53.31	84.23	53.90	83.85	54.49
105	89.24	55.33	88.83	55.98	88.44	56.60	88.04	57.21
110	93.49	57.97	93.06	58.64	92.65	59.29	92.24	59.94
115	97.74	60.60	97.29	61.31	96.86	61.99	96.43	62.66
120	101.99	63.23	101.52	63.97	101.08	64.68	100.62	65.39
125	106.24	65.87	105.75	66.64	105.29	67.38	104.81	68.11
130	110.49	68.50	109.98	69.30	109.50	70.07	109.00	70.83
135	114.74	71.14	114.21	71.97	113.71	72.77	113.20	73.56
140	118.98	73.77	118.44	74.63	117.92	75.46	117.39	76.28
145	123.23	76.41	122.67	77.30	122.13	78.16	121.58	79.01
150	127.48	79.04	126.90	79.96	126.34	80.85	125.78	81.73
155	131.73	81.68	131.14	82.63	130.56	83.55	129.97	84.46
160	135.98	84.31	135.37	85.30	134.77	86.24	134.16	87.18
165	140.23	86.95	139.60	87.97	138.98	88.94	138.35	89.91
170	144.48	89.58	143.83	90.63	143.19	91.63	142.55	92.63
175	148.73	92.22	148.06	93.30	147.40	94.33	146.74	95.35
180	152.98	94.85	152.29	95.96	151.61	97.02	150.93	98.08
185	157.23	97.49	156.52	98.63	155.82	99.72	155.12	100.80
190	161.48	100.12	160.75	101.29	160.04	102.41	159.31	103.53
195	165.73	102.76	164.98	103.96	164.25	105.11	163.51	106.25
200	169.98	105.39	169.21	106.62	168.46	107.80	167.70	108.98
205	174.23	108.03	173.44	109.29	172.67	110.50	171.89	111.70
210	178.48	110.66	177.67	111.96	176.88	113.19	176.09	114.43
215	182.73	113.30	181.90	114.62	181.09	115.89	180.28	117.15
220	186.98	115.93	186.13	117.29	185.31	118.58	184.47	119.87
225	191.23	118.57	190.36	119.95	189.52	121.28	188.67	122.60
230	195.48	121.20	194.59	122.62	193.73	123.97	192.86	125.32
235	199.72	123.83	198.82	125.28	197.94	126.67	197.05	128.05
240	203.97	126.47	203.05	127.95	202.15	129.36	201.24	130.77
245	208.22	129.10	207.28	130.61	206.36	132.06	205.44	133.50
250	212.47	131.74	211.51	133.28	210.57	134.75	209.63	136.22
255	216.72	134.37	215.74	135.95	214.79	137.45	213.82	138.95
260	220.97	137.01	219.97	138.61	219.00	140.14	218.01	141.67
265	225.22	139.64	224.20	141.28	223.21	142.84	222.21	144.39
270	229.47	142.28	228.43	143.94	227.42	145.53	226.40	147.12
275	233.72	144.91	232.66	146.61	231.63	148.23	230.59	149.84
280	237.97	147.55	236.89	149.27	235.84	150.92	234.78	152.57
285	242.22	150.18	241.12	151.94	240.05	153.62	238.98	155.29
290	246.47	152.82	245.35	154.61	244.27	156.31	243.17	158.02
295	250.72	155.45	249.58	157.27	248.48	159.01	247.36	160.74
300	254.97	158.09	253.81	159.94	252.69	161.70	251.55	163.46
350	297.47	184.43	296.13	186.56	294.79	188.67	293.46	190.75

S.D.	66% (33° 25′) H.D.	66% (33° 25′) D.E.	67% (33° 49′) H.D.	67% (33° 49′) D.E.	68% (34° 13′) H.D.	68% (34° 13′) D.E.	69% (34° 36′) H.D.	69% (34° 36′) D.E.
40	33.39	22.03	33.23	22.26	33.08	22.49	32.93	22.71
45	37.56	24.78	37.39	25.04	37.21	25.30	37.04	25.55
50	41.73	27.54	41.54	27.83	41.35	28.12	41.16	28.39
55	45.91	30.29	45.76	30.61	45.48	30.93	45.27	31.23
60	50.08	33.43	49.84	33.39	49.62	33.74	49.38	34.07
65	54.25	35.80	54.00	36.17	53.75	36.55	53.50	36.91
70	58.43	38.55	58.16	38.96	57.88	39,36	57.62	39.75
75	62.60	41.30	62.31	41.74	62.02	42.17	61.74	42.59
80	66.78	44.06	66.47	44.52	66.15	44.99	65.85	45.43
85	70.95	46.81	70.62	47.31	70.29	47.80	69.97	48.27
90	75.12	49.57	74.77	50.09	74.42	50.61	74.08	51.11
95	79.30	52.32	78.93	52.87	78.56	53.42	78.20	53.95
100	83.47	55.07	83.08	55.65	82.69	56.23	82.31	56.78
105	87.64	57.83	87.24	58.44	86.83	59.04	86.43	59.62
110	91.82	60.58	91.39	61.22	90.96	61.86	90.55	62.46
115	95.99	63.33	95.54	64.00	95.10	64.67	94.66	65.30
120	100.16	66.09	99.70	66.78	99.23	67.48	98.78	68.14
125	104.34	68.84	103.85	69.57	103.36	70.29	102.89	70.98
130	108.51	71.59	108.00	72.35	107.50	73.10	107.01	73.82
135	112.68	74.35	112.16	75.13	111.63	75.91	111.12	76.66
140	116.87	77.10	116.32	77.92	115.77	78.73	115.24	79.50
145	121.03	79.85	120.47	80.70	119.90	81.54	119.35	82.34
150	125.20	82.61	124.62	83.48	124.04	84.35	123.47	85.18
155	129.38	85.36	128.78	86.26	128.17	87.16	127.59	88.02
160	133.55	88.12	132.93	89.05	132.31	89.97	131.70	90.85
165	137.72	90.87	137.09	91.83	136.44	92.78	135.82	93.69
170	141.90	93.62	141.24	94.61	140.58	95.60	139.93	96.53
175	146.07	96.38	145.39	97.39	144.71	98.41	144.05	99.37
180	150.24	99.13	149.55	100.18	148.85	101.22	148.16	102.21
185	154.42	101.88	153.70	102.96	152.98	104.03	152.28	105.05
190	158.59	104.64	157.86	105.74	157.11	106.84	156.40	107.89
195	162.76	107.39	162.01	108.52	161.25	109.65	160.51	110.73
200	166.94	110.14	166.16	111.31	165.38	112.46	164.63	113.57
205	171.11	112.90	170.32	114.09	169.52	115.28	168.74	116.41
210	175.28	115.65	174.47	116.87	173.65	118.09	172.89	119.25
215	179.46	118.41	178.63	119.66	177.79	120.90	176.97	122.09
220	183.63	121.16	182.78	122.44	181.92	123.71	181.09	124.93
225	187.80	123.91	186.94	125.22	186.06	126.52	185.21	127.76
230	191.98	126.67	191.09	128.00	190.19	129.33	189.32	130.60
235	196.15	129.42	195.24	130.79	194.33	132.15	193.44	133.44
240	200.33	132.17	199.40	133.57	198.46	134.96	197.55	136.28
245	204.50	134.93	203.55	136.35	202.59	137.77	201.67	139.12
250	208.67	137.68	207.71	139.13	206.73	140.58	205.78	141.96
255	212.85	140.43	211.86	141.92	210.86	143.39	209.90	144.80
260	217.02	143.19	216.01	144.70	215.00	146.20	214.02	147.64
265	221.19	145.94	220.17	147.48	219.13	149.02	218.13	150.48
270	225.37	148.70	224.32	150.27	223.27	151.83	222.25	153.32
275	229.54	151.45	228.48	153.05	227.40	154.64	226.36	156.16
280	233.71	154.20	232.63	155.83	231.54	157.45	230.49	159.00
285	237.88	156.96	236.78	158.61	235.67	160.26	234.59	161.84
290	242.06	159.71	240.94	161.40	239.81	163.07	238.71	164.67
295	246.23	162.46	245.09	164.18	243.94	165.89	242.83	167.51
300	250.41	165.22	249.25	166.96	248.08	168.70	246.94	170.35
350	292.11	192.79	290.77	194.82	289.42	196.81	288.08	198.77

S.D.	70% (35° 00′) H.D.	D.E.	71% (35° 22′) H.D.	D.E.	72% (35° 45′) H.D.	D.E.	73% (36° 08′) H.D.	D.E.
40	32.77	22.94	32.62	23.15	32.46	23.37	32.31	23.59
45	36.86	25.81	36.70	26.05	36.52	26.28	36.34	26.53
50	40.96	28.68	40.77	28.94	40.58	29.21	40.38	29.48
55	45.05	31.55	44.85	31.83	44.63	32.13	44.42	32.43
60	49.15	34.41	48.93	34.73	48.69	35.05	48.46	35.38
65	53.24	37.28	53.01	37.62	52.75	37.97	52.50	38.33
70	57.34	40.15	57.08	40.52	56.81	40.90	56.54	41.28
75	61.44	43.02	61.16	43.41	60.86	43.81	60.57	44.22
80	65.53	45.89	65.24	46.30	64.93	46.74	64.61	47.17
85	69.63	48.75	69.31	49.20	68.98	49.66	68.65	50.12
90	73.72	51.62	73.39	52.09	73.04	52.58	72.69	53.07
95	77.82	54.49	77.47	54.99	77.09	55.50	76.73	56.02
100	81.92	57.36	81.55	57.88	81.16	58.42	80.76	58.97
105	86.01	60.23	85.62	60.77	85.21	61.34	84.80	61.91
110	90.11	63.09	89.70	63.67	89.27	64.27	88.84	64.86
115	94.20	65.96	93.78	66.56	93.32	67.18	92.88	67.81
120	98.30	68.83	97.86	69.46	97.39	70.11	96.92	70.76
125	102.39	71.70	101.93	72.35	101.44	73.02	100.96	73.71
130	106.49	74.56	106.01	75.24	105.50	75.95	104.99	76.66
135	110.59	77.43	110.09	78.14	109.55	78.87	109.03	79.60
140	114.68	80.30	114.17	81.03	113.62	81.79	113.07	82.55
145	118.78	83.17	118.24	83.93	117.67	84.71	117.11	85.50
150	122.87	86.04	122.32	86.82	121.74	87.64	121.15	88.45
155	126.97	88.90	126.40	89.72	125.78	90.55	125.19	91.40
160	131.06	91.77	130.47	92.61	129.85	93.48	129.22	94.35
165	135.16	94.64	134.55	95.50	133.90	96.39	133.26	97.29
170	139.26	97.51	138.63	98.40	137.97	99.32	137.30	100.24
175	143.35	100.38	142.71	101.29	142.01	102.23	141.34	103.19
180	147.45	103.24	146.78	104.19	146.09	105.16	145.38	106.14
185	151.54	106.11	150.86	107.08	150.13	108.08	149.41	109.09
190	155.64	108.98	154.94	109.97	154.20	111.01	153.45	112.04
195	159.73	111.85	159.06	112.87	158.24	113.92	157.49	114.98
200	163.83	114.72	163.09	115.76	162.31	116.85	161.53	117.93
205	167.93	117.58	167.17	118.63	166.36	119.76	165.57	120.88
210	172.02	120.45	171.25	121.55	170.43	122.69	169.61	123.83
215	176.12	123.32	175.32	124.44	174.47	125.60	173.64	126.78
220	180.21	126.19	179.40	127.34	178.55	128.53	177.68	129.73
225	184.31	129.05	183.48	130.23	182.59	131.44	181.72	132.67
230	188.40	131.92	187.56	133.13	186.66	134.38	185.76	135.62
235	192.50	134.79	191.63	136.02	190.70	137.29	189.80	138.57
240	196.60	137.66	195.71	138.91	194.78	140.22	193.84	141.52
245	200.69	140.53	199.79	141.81	198.82	143.13	197.87	144.47
250	204.79	143.39	203.87	144.70	202.89	146.06	201.91	147.42
255	208.88	146.26	207.94	147.60	206.93	148.97	205.95	150.36
260	212.98	149.13	212.02	150.49	211.01	151.90	209.99	153.31
265	217.08	152.00	216.10	153.38	215.05	154.81	214.03	156.26
270	221.17	154.87	220.18	156.28	219.12	157.75	218.06	159.21
275	225.27	157.73	224.25	159.17	223.16	160.66	222.10	162.16
280	229.36	160.60	228.33	162.07	227.24	163.59	226.14	165.11
285	233.46	163.47	232.41	164.96	231.28	166.50	230.18	168.05
290	237.55	166.34	236.48	167.85	235.34	169.42	234.22	171.00
295	241.65	169.21	240.56	170.75	239.39	172.34	238.26	173.95
300	245.75	172.07	244.64	173.64	243.45	175.26	242.29	176.90
350	286.73	200.71	285.38	202.62	284.04	204.51	282.69	206.36

S.D.	74% (36° 30′) H.D.	D.E.	75% (36° 52′) H.D.	D.E.	76% (37° 14′) H.D.	D.E.	77% (37° 36′) H.D.	D.E.
40	32.15	23.79	32.00	24.00	31.85	24.20	31.69	24.41
45	36.17	26.77	36.00	27.00	35.83	27.23	35.65	27.46
50	40.19	29.74	40.00	30.00	39.81	30.25	39.61	30.51
55	44.21	32.72	44.00	33.00	43.79	33.28	43.58	33.56
60	48.23	35.69	48.00	36.00	47.77	36.30	47.54	36.61
65	52.25	38.66	52.00	39.00	51.75	39.33	51.50	39.66
70	56.27	41.64	56.00	42.00	55.73	42.35	55.46	42.71
75	60.29	44.61	60.00	45.00	59.71	45.38	59.42	45.76
80	64.31	47.59	64.00	48.00	63.69	48.40	63.38	48.81
85	68.33	50.56	68.00	51.00	67.68	51.43	67.34	51.86
90	72.35	53.53	72.00	54.00	71.66	54.46	71.31	54.91
95	76.37	56.51	76.00	57.00	75.64	57.48	75.27	57.96
100	80.39	59.48	80.00	60.00	79.62	60.51	79.23	61.01
105	84.40	62.46	84.00	63.00	83.60	63.53	83.19	64.06
110	88.42	65.43	88.00	66.00	87.58	66.56	87.15	67.11
115	92.44	68.40	92.00	68.99	91.56	69.58	91.11	70.17
120	96.46	71.38	96.00	71.99	95.54	72.61	95.07	73.22
125	100.48	74.35	100.00	74.99	99.52	75.63	99.04	76.27
130	104.50	77.33	104.00	77.99	103.50	78.66	103.00	79.32
135	108.52	80.30	108.00	80.99	107.48	81.69	106.96	82.37
140	112.54	83.28	112.00	83.99	111.46	84.71	110.92	85.42
145	116.56	86.25	116.00	86.99	115.45	87.73	114.88	88.47
150	120.59	89.22	120.01	89.99	119.43	90.76	118.84	91.52
155	124.60	92.20	124.01	92.99	123.41	93.78	122.80	94.57
160	128.62	95.17	128.01	95.99	127.39	96.81	126.77	97.62
165	132.64	98.15	132.01	98.99	131.37	99.84	130.73	100.67
170	136.66	101.12	136.01	101.99	135.35	102.86	134.69	103.72
175	140.67	104.09	140.01	104.99	139.33	105.89	138.65	106.77
180	144.69	107.07	144.01	107.99	143.31	108.91	142.61	109.82
185	148.71	110.04	148.01	110.99	147.29	111.94	146.57	112.88
190	152.73	113.02	152.01	113.99	151.27	114.96	150.54	115.93
195	156.75	115.99	156.01	116.99	155.25	117.99	154.50	118.98
200	160.77	118.96	160.01	119.99	159.24	121.01	158.46	122.03
205	164.79	121.94	164.01	122.99	163.22	124.04	162.42	125.08
210	168.81	124.91	168.01	125.99	167.20	127.06	166.38	128.13
215	172.83	127.89	172.01	128.99	171.18	130.09	170.34	131.18
220	176.85	130.86	176.01	131.99	175.16	133.11	174.30	134.23
225	180.87	133.84	180.01	134.99	179.14	136.14	178.27	137.28
230	184.89	136.81	184.01	137.99	183.12	139.16	182.23	140.33
235	188.91	139.78	188.01	140.99	187.10	142.19	186.19	143.38
240	192.93	142.76	192.01	143.99	191.08	145.21	190.15	146.43
245	196.94	145.73	196.01	146.99	195.06	148.24	194.11	149.48
250	200.96	148.71	200.01	149.99	199.04	151.27	198.07	152.53
255	204.98	151.68	204.01	152.99	203.03	154.29	202.03	155.58
260	209.00	154.65	208.01	155.99	207.01	157.32	206.00	158.64
265	213.02	157.63	212.01	158.99	210.99	160.34	209.96	161.69
270	217.04	160.60	216.01	161.99	214.97	163.37	213.92	164.74
275	221.06	163.58	220.01	164.99	218.95	166.39	217.88	167.79
280	225.08	166.55	224.01	167.99	222.93	169.42	221.84	170.84
285	229.10	169.52	228.01	170.99	226.92	172.44	225.80	173.89
290	233.12	172.50	232.01	173.99	230.89	175.47	229.76	176.94
295	237.14	175.47	236.01	176.99	234.87	178.49	233.73	179.99
300	241.16	178.45	240.01	179.99	238.85	181.52	237.69	183.04
350	281.34	208.20	280.00	210.00	278.66	211.78	277.32	213.53

	78% (37° 57′)		79% (38° 19′)		80% (38° 40′)		81% (39° 00′)	
S.D.	H.D.	D.E.	H.D.	D.E.	H.D.	D.E.	H.D.	D.E.
40	31.54	24.60	31.38	24.80	31.23	24.99	31.09	25.17
45	35.48	27.67	35.31	27.90	35.13	28.12	34.97	28.32
50	39.43	30.75	39.23	31.00	39.04	31.24	38.86	31.47
55	43.37	33.82	43.15	34.10	42.94	34.36	42.74	34.61
60	47.31	36.90	47.08	37.20	46.85	37.49	46.63	37.76
65	51.26	39.97	51.00	40.30	50.75	40.61	50.51	40.91
70	55.20	43.05	54.92	43.40	54.66	43.74	54.40	44.05
75	59.14	46.12	58.84	46.50	58.56	46.86	58.29	47.20
80	63.08	49.20	62.77	49.60	62.46	49.98	62.17	50.35
85	67.03	52.27	66.69	52.70	66.37	53.11	66.06	53.49
90	70.97	55.35	70.61	55.80	70.27	56.23	69.94	56.64
95	74.91	58.42	74.54	58.90	74.18	59.35	73.83	59.79
100	78.85	61.50	78.46	62.00	78.08	62.48	77.71	62.93
105	82.80	64.57	82.38	65.10	81.98	65.60	81.60	66.08
110	86.74	67.65	86.31	68.20	85.89	68.73	85.49	69.23
115	90.68	70.72	90.23	71.30	89.79	71.85	89.37	72.37
120	94.63	73.80	94.15	74.40	93.70	74.97	93.26	75.52
125	98.57	76.87	98.07	77.50	97.60	78.10	97.14	78.67
130	102.51	79.95	102.00	80.60	101.50	81.22	101.03	81.81
135	106.45	83.02	105.92	83.70	105.41	84.35	104.91	84.96
140	110.40	86.10	109.84	86.80	109.31	87.47	108.80	88.10
145	114.34	89.17	113.77	89.90	113.21	90.59	112.69	91.25
150	118.28	92.25	117.69	93.00	117.11	93.72	116.57	94.40
155	122.22	95.32	121.61	96.10	121.02	96.84	120.46	97.54
160	126.17	98.40	125.54	99.20	124.93	99.97	124.34	100.69
165	130.11	101.47	129.46	102.30	128.83	103.09	128.23	103.84
170	134.05	104.55	133.38	105.40	132.73	106.21	132.11	106.98
175	138.00	107.62	137.30	108.50	136.64	109.34	136.00	110.13
180	141.94	110.70	141.23	111.60	140.54	112.46	139.89	113.28
185	145.88	113.77	145.15	114.70	144.45	115.59	143.77	116.42
190	149.82	116.84	149.07	117.80	148.35	118.71	147.66	119.57
195	153.77	119.92	153.00	120.90	152.25	121.83	151.54	122.72
200	157.71	122.99	156.92	124.00	156.16	124.96	155.43	125.86
205	161.65	126.07	160.84	127.10	160.06	128.08	159.31	129.01
210	165.60	129.14	164.77	130.20	163.97	131.21	163.20	132.16
215	169.54	132.22	168.69	133.30	167.87	134.33	167.09	135.30
220	173.48	135.29	172.61	136.40	171.77	137.45	170.97	138.45
225	177.42	138.37	176.53	139.50	175.68	140.58	174.86	141.60
230	181.37	141.44	180.46	142.60	179.58	143.70	178.74	144.74
235	185.31	144.52	184.38	145.70	183.49	146.83	182.63	147.89
240	189.25	147.59	188.30	148.80	187.39	149.95	186.52	151.04
245	193.19	150.67	192.23	151.90	191.29	153.07	190.40	154.18
250	197.14	153.74	196.15	155.00	195.20	156.20	194.29	157.33
255	201.08	156.82	200.07	158.10	199.10	159.32	198.17	160.48
260	205.02	159.89	203.99	161.20	203.01	162.45	202.06	163.62
265	208.97	162.97	207.92	164.30	206.91	165.57	205.94	166.77
270	212.91	166.04	211.84	167.40	210.81	168.69	209.83	169.92
275	216.85	169.12	215.76	170.50	214.72	171.82	213.72	173.06
280	220.79	172.19	219.69	173.60	218.62	174.94	217.60	176.21
285	224.74	175.27	223.61	176.70	222.53	178.06	221.49	179.36
290	228.68	178.34	227.53	179.80	226.43	181.19	225.37	182.50
295	232.62	181.42	231.46	182.90	230.33	184.31	229.26	185.65
300	236.56	184.49	235.38	186.00	234.24	187.44	233.14	188.80
350	275.98	215.26	274.64	216.96	273.30	218.04	271.97	220.30

%	Angle	Cos	Sin	%	Angle	Cos	Sin
82	39°21′	.7733	.6341	116	49°14′	.6529	.7574
83	39°42′	.7695	.6387	117	49°29′	.6497	.7602
84	40°02′	.7657	.6432	118	49°43′	.6465	.7629
85	40°22′	.7619	.6476	119	49°58′	.6433	.7656
86	40°42′	.7582	.6520	120	50°12′	.6402	.7682
87	41°01′	.7544	.6564	121	50°26′	.6370	.7708
88	41°21′	.7507	.6606	122	50°40′	.6339	.7734
89	41°40′	.7470	.6648	123	50°53′	.6308	.7759
90	41°59′	.7433	.6690	124	51°07′	.6278	.7784
91	42°18′	.7396	.6730	125	51°20′	.6247	.7809
92	42°37′	.7359	.6771	126	51°34′	.6217	.7833
93	42°55′	.7323	.6810	127	51°47′	.6186	.7857
94	43°14′	.7286	.6849	128	52°00′	.6156	.7880
95	43°32′	.7250	.6887	129	52°13′	.6127	.7903
96	43°50′	.7214	.6925	130	52°26′	.6097	.7926
97	44°08′	.7178	.6963	131	52°39′	.6068	.7949
98	44°25′	.7142	.6999	132	52°51′	.6039	.7971
99	44°43′	.7107	.7035	133	53°04′	.6010	.7993
100	45°00′	.7071	.7071	134	53°16′	.5981	.8014
101	45°17′	.7036	.7106	135	53°28′	.5952	.8036
102	45°34′	.7001	.7141	136	53°40′	.5924	.8057
103	45°51′	.6966	.7175	137	53°52′	.5896	.8077
104	46°07′	.6931	.7208	138	54°04′	.5868	.8097
105	46°24′	.6897	.7241	139	54°16′	.5840	.8118
106	46°40′	.6862	.7274	140	54°28′	.5812	.8137
107	46°56′	.6828	.7306	141	54°39′	.5785	.8157
108	47°12′	.6794	.7338	142	54°51′	.5750	.8176
109	47°28′	.6760	.7368	143	55°02′	.5731	.8195
110	47°44′	.6727	.7399	144	55°13′	.5704	.8214
111	47°59′	.6693	.7430	145	55°24′	.5677	.8232
112	48°14′	.6660	.7459	146	55°35′	.5651	.8250
113	48°30′	.6627	.7489	147	55°46′	.5625	.8268
114	48°45′	.6594	.7518	148	55°57′	.5599	.8286
115	48°59′	.6562	.7546	149	56°08′	.5573	.8303
				150	56°19′	.5547	.8321

Table 8. Natural trigonometric functions.

Values of the trigonometric functions of angles for each minute from 0°–360°. For degrees indicated at the top of the page use the column headings at the top. For degrees indicated at the bottom use the column indications at the bottom.

With degrees at the left of each block (top or bottom), use the minute column at the left and with degrees at the right of each block use the minute column at the right.

These tables may also be used to compute latitudes and departures of courses when the distance and azimuth are known. Use the sine and cosine values of the angular value of the azimuth as directed above and then apply the correct sign depending in which quadrant the azimuth lies. For example the angular values of 32° and 57° are shown in the tables as follows:

32° (212°)	(327°) 147°
122° (302°)	(237°) 57°

Using the proper trigonometric value to compute the latitudes and departures, the following signs would be applied to the computed values.

32° (lat.+, dep.+), (212°)(lat.–, dep.–)
147° (lat.–, dep.+), (327°)(lat.+, dep.–)
57° (lat.+, dep.+), (237°)(lat.–, dep.–)
122° (lat.–, dep.+), (302°)(lat.+, dep.–)

NATURAL TRIGONOMETRIC FUNCTIONS

0° (180°) (359°) **179°**

′	Sin	Tan	Cot	Cos	′
0	.00000	.00000		1.0000	**60**
1	.00029	.00029	3437.7	1.0000	59
2	.00058	.00058	1718.9	1.0000	58
3	.00087	.00087	1145.9	1.0000	57
4	.00116	.00116	859.44	1.0000	56
5	.00145	.00145	687.55	1.0000	**55**
6	.00175	.00175	572.96	1.0000	54
7	.00204	.00204	491.11	1.0000	53
8	.00233	.00233	429.72	1.0000	52
9	.00262	.00262	381.97	1.0000	51
10	.00291	.00291	343.77	1.0000	**50**
11	.00320	.00320	312.52	.99999	49
12	.00349	.00349	286.48	.99999	48
13	.00378	.00378	264.44	.99999	47
14	.00407	.00407	245.55	.99999	46
15	.00436	.00436	229.18	.99999	**45**
16	.00465	.00465	214.86	.99999	44
17	.00495	.00495	202.22	.99999	43
18	.00524	.00524	190.98	.99999	42
19	.00553	.00553	180.93	.99998	41
20	.00582	.00582	171.89	.99998	**40**
21	.00611	.00611	163.70	.99998	39
22	.00640	.00640	156.26	.99998	38
23	.00669	.00669	149.47	.99998	37
24	.00698	.00698	143.24	.99998	36
25	.00727	.00727	137.51	.99997	**35**
26	.00756	.00756	132.22	.99997	34
27	.00785	.00785	127.32	.99997	33
28	.00814	.00815	122.77	.99997	32
29	.00844	.00844	118.54	.99996	31
30	.00873	.00873	114.59	.99996	**30**
31	.00902	.00902	110.89	.99996	29
32	.00931	.00931	107.43	.99996	28
33	.00960	.00960	104.17	.99995	27
34	.00989	.00989	101.11	.99995	26
35	.01018	.01018	98.218	.99995	**25**
36	.01047	.01047	95.489	.99995	24
37	.01076	.01076	92.908	.99994	23
38	.01105	.01105	90.463	.99994	22
39	.01134	.01135	88.144	.99994	21
40	.01164	.01164	85.940	.99993	**20**
41	.01193	.01193	83.844	.99993	19
42	.01222	.01222	81.847	.99993	18
43	.01251	.01251	79.943	.99992	17
44	.01280	.01280	78.126	.99992	16
45	.01309	.01309	76.390	.99991	**15**
46	.01338	.01338	74.729	.99991	14
47	.01367	.01367	73.139	.99991	13
48	.01396	.01396	71.615	.99990	12
49	.01425	.01425	70.153	.99990	11
50	.01454	.01455	68.750	.99989	**10**
51	.01483	.01484	67.402	.99989	9
52	.01513	.01513	66.105	.99989	8
53	.01542	.01542	64.858	.99988	7
54	.01571	.01571	63.657	.99988	6
55	.01600	.01600	62.499	.99987	**5**
56	.01629	.01629	61.383	.99987	4
57	.01658	.01658	60.306	.99986	3
58	.01687	.01687	59.266	.99986	2
59	.01716	.01716	58.261	.99985	1
60	.01745	.01746	57.290	.99985	**0**
′	Cos	Cot	Tan	Sin	′

90° (270°) (269°) **89°**

1° (181°) (358°) **178°**

′	Sin	Tan	Cot	Cos	′
0	.01745	.01746	57.290	.99985	**60**
1	.01774	.01775	56.351	.99984	59
2	.01803	.01804	55.442	.99984	58
3	.01832	.01833	54.561	.99983	57
4	.01862	.01862	53.709	.99983	56
5	.01891	.01891	52.882	.99982	**55**
6	.01920	.01920	52.081	.99982	54
7	.01949	.01949	51.303	.99981	53
8	.01978	.01978	50.549	.99980	52
9	.02007	.02007	49.816	.99980	51
10	.02036	.02036	49.104	.99979	**50**
11	.02065	.02066	48.412	.99979	49
12	.02094	.02095	47.740	.99978	48
13	.02123	.02124	47.085	.99977	47
14	.02152	.02153	46.449	.99977	46
15	.02181	.02182	45.829	.99976	**45**
16	.02211	.02211	45.226	.99976	44
17	.02240	.02240	44.639	.99975	43
18	.02269	.02269	44.066	.99974	42
19	.02298	.02298	43.508	.99974	41
20	.02327	.02328	42.964	.99973	**40**
21	.02356	.02357	42.433	.99972	39
22	.02385	.02386	41.916	.99972	38
23	.02414	.02415	41.411	.99971	37
24	.02443	.02444	40.917	99970	36
25	.02472	.02473	40.436	.99969	**35**
26	.02501	.02502	39.965	.99969	34
27	.02530	.02531	39.506	.99968	33
28	.02560	.02560	39.057	.99967	32
29	.02589	.02589	38.618	.99966	31
30	.02618	.02619	38.188	.99966	**30**
31	.02647	.02648	37.769	.99965	29
32	.02676	.02677	37.358	.99964	28
33	.02705	.02706	36.956	.99963	27
34	.02734	.02735	36.563	.99963	26
35	.02763	.02764	36.178	.99962	**25**
36	.02792	.02793	35.801	.99961	24
37	.02821	.02822	35.431	.99960	23
38	.02850	.02851	35.070	.99959	22
39	.02879	.02881	34.715	.99959	21
40	.02908	.02910	34.368	.99958	**20**
41	.02938	.02939	34.027	.99957	19
42	.02967	.02968	33.694	.99956	18
43	.02996	.02997	33.366	.99955	17
44	.03025	.03026	33.045	.99954	16
45	.03054	.03055	32.730	.99953	**15**
46	.03083	.03084	32.421	.99952	14
47	.03112	.03114	32.118	.99952	13
48	.03141	.03143	31.821	.99951	12
49	.03170	.03172	31.528	.99950	11
50	.03199	.03201	31.242	.99949	**10**
51	.03228	.03230	30.960	.99948	9
52	.03257	.03259	30.683	.99947	8
53	.03286	.03288	30.412	.99946	7
54	.03316	.03317	30.145	.99945	6
55	.03345	.03346	29.882	.99944	**5**
56	.03374	.03376	29.624	.99943	4
57	.03403	.03405	29.371	.99942	3
58	.03432	.03434	29.122	.99941	2
59	.03461	.03463	28.877	.99940	1
60	.03490	.03492	28.636	.99939	**0**
′	Cos	Cot	Tan	Sin	′

91° (271°) (268°) **88°**

NATURAL TRIGONOMETRIC FUNCTIONS

2° (182°) (357°) 177°

′	Sin	Tan	Cot	Cos	′
0	.03490	.03492	28.636	.99939	60
1	.03519	.03521	28.399	.99938	59
2	.03548	.03550	28.166	.99937	58
3	.03577	.03579	27.937	.99936	57
4	.03606	.03609	27.712	.99935	56
5	.03635	.03638	27.490	.99934	55
6	.03664	.03667	27.271	.99933	54
7	.03693	.03696	27.057	.99932	53
8	.03723	.03725	26.845	.99931	52
9	.03752	.03754	26.637	.99930	51
10	.03781	.03783	26.432	.99929	50
11	.03810	.03812	26.230	.99927	49
12	.03839	.03842	26.031	.99926	48
13	.03868	.03871	25.835	.99925	47
14	.03897	.03900	25.642	.99924	46
15	.03926	.03929	25.452	.99923	45
16	.03955	.03958	25.264	.99922	44
17	.03984	.03987	25.080	.99921	43
18	.04013	.04016	24.898	.99919	42
19	.04042	.04046	24.719	.99918	41
20	.04071	.04075	24.542	.99917	40
21	.04100	.04104	24.368	.99916	39
22	.04129	.04133	24.196	.99915	38
23	.04159	.04162	24.026	.99913	37
24	.04188	.04191	23.859	.99912	36
25	.04217	.04220	23.695	.99911	35
26	.04246	.04250	23.532	.99910	34
27	.04275	.04279	23.372	.99909	33
28	.04304	.04308	23.214	.99907	32
29	.04333	.04337	23.058	.99906	31
30	.04362	.04366	22.904	.99905	30
31	.04391	.04395	22.752	.99904	29
32	.04420	.04424	22.602	.99902	28
33	.04449	.04454	22.454	.99901	27
34	.04478	.04483	22.308	.99900	26
35	.04507	.04512	22.164	.99898	25
36	.04536	.04541	22.022	.99897	24
37	.04565	.04570	21.881	.99896	23
38	.04594	.04599	21.743	.99894	22
39	.04623	.04628	21.606	.99893	21
40	.04653	.04658	21.470	.99892	20
41	.04682	.04687	21.337	.99890	19
42	.04711	.04716	21.205	.99889	18
43	.04740	.04745	21.075	.99888	17
44	.04769	.04774	20.946	.99886	16
45	.04798	.04803	20.819	.99885	15
46	.04827	.04833	20.693	.99883	14
47	.04856	.04862	20.569	.99882	13
48	.04885	.04891	20.446	.99881	12
49	.04914	.04920	20.325	.99879	11
50	.04943	.04949	20.206	.99878	10
51	.04972	.04978	20.087	.99876	9
52	.05001	.05007	19.970	.99875	8
53	.05030	.05037	19.855	.99873	7
54	.05059	.05066	19.740	.99872	6
55	.05088	.05095	19.627	.99870	5
56	.05117	.05124	19.516	.99869	4
57	.05146	.05153	19.405	.99867	3
58	.05175	.05182	19.296	.99866	2
59	.05205	.05212	19.188	.99864	1
60	.05234	.05241	19.081	.99863	0
′	Cos	Cot	Tan	Sin	′

92° (272°) (267°) 87°

3° (183°) (356°) 176°

′	Sin	Tan	Cot	Cos	′
0	.05234	.05241	19.081	.99863	60
1	.05263	.05270	18.976	.99861	59
2	.05292	.05299	18.871	.99860	58
3	.05321	.05328	18.768	.99858	57
4	.05350	.05357	18.666	.99857	56
5	.05379	.05387	18.564	.99855	55
6	.05408	.05416	18.464	.99854	54
7	.05437	.05445	18.366	.99852	53
8	.05466	.05474	18.268	.99851	52
9	.05495	.05503	18.171	.99849	51
10	.05524	.05533	18.075	.99847	50
11	.05553	.05562	17.980	.99846	49
12	.05582	.05591	17.886	.99844	48
13	.05611	.05620	17.793	.99842	47
14	.05640	.05649	17.702	.99841	46
15	.05669	.05678	17.611	.99839	45
16	.05698	.05708	17.521	.99838	44
17	.05727	.05737	17.431	.99836	43
18	.05756	.05766	17.343	.99834	42
19	.05785	.05795	17.256	.99833	41
20	.05814	.05824	17.169	.99831	40
21	.05844	.05854	17.084	.99829	39
22	.05873	.05883	16.999	.99827	38
23	.05902	.05912	16.915	.99826	37
24	.05931	.05941	16.832	.99824	36
25	.05960	.05970	16.750	.99822	35
26	.05989	.05999	16.668	.99821	34
27	.06018	.06029	16.587	.99819	33
28	.06047	.06058	16.507	.99817	32
29	.06076	.06087	16.428	.99815	31
30	.06105	.06116	16.350	.99813	30
31	.06134	.06145	16.272	.99812	29
32	.06163	.06175	16.195	.99810	28
33	.06192	.06204	16.119	.99808	27
34	.06221	.06233	16.043	.99806	26
35	.06250	.06262	15.969	.99804	25
36	.06279	.06291	15.895	.99803	24
37	.06308	.06321	15.821	.99801	23
38	.06337	.06350	15.748	.99799	22
39	.06366	.06379	15.676	.99797	21
40	.06395	.06408	15.605	.99795	20
41	.06424	.06438	15.534	.99793	19
42	.06453	.06467	15.464	.99792	18
43	.06482	.06496	15.394	.99790	17
44	.06511	.06525	15.325	.99788	16
45	.06540	.06554	15.257	.99786	15
46	.06569	.06584	15.189	.99784	14
47	.06598	.06613	15.122	.99782	13
48	.06627	.06642	15.056	.99780	12
49	.06656	.06671	14.990	.99778	11
50	.06685	.06700	14.924	.99776	10
51	.06714	.06730	14.860	.99774	9
52	.06743	.06759	14.795	.99772	8
53	.06773	.06788	14.732	.99770	7
54	.06802	.06817	14.669	.99768	6
55	.06831	.06847	14.606	.99766	5
56	.06860	.06876	14.544	.99764	4
57	.06889	.06905	14.482	.99762	3
58	.06918	.06934	14.421	.99760	2
59	.06947	.06963	14.361	.99758	1
60	.06976	.06993	14.301	.99756	0
′	Cos	Cot	Tan	Sin	′

93° (273°) (266°) 86°

NATURAL TRIGONOMETRIC FUNCTIONS

4° (184°) — **(355°) 175°**

′	Sin	Tan	Cot	Cos	′
0	.06976	.06993	14.301	.99756	60
1	.07005	.07022	14.241	.99754	59
2	.07034	.07051	14.182	.99752	58
3	.07063	.07080	14.124	.99750	57
4	.07092	.07110	14.065	.99748	56
5	.07121	.07139	14.008	.99746	55
6	.07150	.07168	13.951	.99744	54
7	.07179	.07197	13.894	.99742	53
8	.07208	.07227	13.838	.99740	52
9	.07237	.07256	13.782	.99738	51
10	.07266	.07285	13.727	.99736	50
11	.07295	.07314	13.672	.99734	49
12	.07324	.07344	13.617	.99731	48
13	.07353	.07373	13.563	.99729	47
14	.07382	.07402	13.510	.99727	46
15	.07411	.07431	13.457	.99725	45
16	.07440	.07461	13.404	.99723	44
17	.07469	.07490	13.352	.99721	43
18	.07498	.07519	13.300	.99719	42
19	.07527	.07548	13.248	.99716	41
20	.07556	.07578	13.197	.99714	40
21	.07585	.07607	13.146	.99712	39
22	.07614	.07636	13.096	.99710	38
23	.07643	.07665	13.046	.99708	37
24	.07672	.07695	12.996	.99705	36
25	.07701	.07724	12.947	.99703	35
26	.07730	.07753	12.898	.99701	34
27	.07759	.07782	12.850	.99699	33
28	.07788	.07812	12.801	.99696	32
29	.07817	.07841	12.754	.99694	31
30	.07846	.07870	12.706	.99692	30
31	.07875	.07899	12.659	.99689	29
32	.07904	.07929	12.612	.99687	28
33	.07933	.07958	12.566	.99685	27
34	.07962	.07987	12.520	.99683	26
35	.07991	.08017	12.474	.99680	25
36	.08020	.08046	12.429	.99678	24
37	.08049	.08075	12.384	.99676	23
38	.08078	.08104	12.339	.99673	22
39	.08107	.08134	12.295	.99671	21
40	.08136	.08163	12.251	.99668	20
41	.08165	.08192	12.207	.99666	19
42	.08194	.08221	12.163	.99664	18
43	.08223	.08251	12.120	.99661	17
44	.08252	.08280	12.077	.99659	16
45	.08281	.08309	12.035	.99657	15
46	.08310	.08339	11.992	.99654	14
47	.08339	.08368	11.950	.99652	13
48	.08368	.08397	11.909	.99649	12
49	.08397	.08427	11.867	.99647	11
50	.08426	.08456	11.826	.99644	10
51	.08455	.08485	11.785	.99642	9
52	.08484	.08514	11.745	.99639	8
53	.08513	.08544	11.705	.99637	7
54	.08542	.08573	11.664	.99635	6
55	.08571	.08602	11.625	.99632	5
56	.08600	.08632	11.585	.99630	4
57	.08629	.08661	11.546	.99627	3
58	.08658	.08690	11.507	.99625	2
59	.08687	.08720	11.468	.99622	1
60	.08716	.08749	11.430	.99619	0
′	Cos	Cot	Tan	Sin	′

94° (274°) — **(265°) 85°**

5° (185°) — **(354°) 174°**

′	Sin	Tan	Cot	Cos	′
0	.08716	.08749	11.430	.99619	60
1	.08745	.08778	11.392	.99617	59
2	.08774	.08807	11.354	.99614	58
3	.08803	.08837	11.316	.99612	57
4	.08831	.08866	11.279	.99609	56
5	.08860	.08895	11.242	.99607	55
6	.08889	.08925	11.205	.99604	54
7	.08918	.08954	11.168	.99602	53
8	.08947	.08983	11.132	.99599	52
9	.08976	.09013	11.095	.99596	51
10	.09005	.09042	11.059	.99594	50
11	.09034	.09071	11.024	.99591	49
12	.09063	.09101	10.988	.99588	48
13	.09092	.09130	10.953	.99586	47
14	.09121	.09159	10.918	.99583	46
15	.09150	.09189	10.883	.99580	45
16	.09179	.09218	10.848	.99578	44
17	.09208	.09247	10.814	.99575	43
18	.09237	.09277	10.780	.99572	42
19	.09266	.09306	10.746	.99570	41
20	.09295	.09335	10.712	.99567	40
21	.09324	.09365	10.678	.99564	39
22	.09353	.09394	10.645	.99562	38
23	.09382	.09423	10.612	.99559	37
24	.09411	.09453	10.579	.99556	36
25	.09440	.09482	10.546	.99553	35
26	.09469	.09511	10.514	.99551	34
27	.09498	.09541	10.481	.99548	33
28	.09527	.09570	10.449	.99545	32
29	.09556	.09600	10.417	.99542	31
30	.09585	.09629	10.385	.99540	30
31	.09614	.09658	10.354	.99537	29
32	.09642	.09688	10.322	.99534	28
33	.09671	.09717	10.291	.99531	27
34	.09700	.09746	10.260	.99528	26
35	.09729	.09776	10.229	.99526	25
36	.09758	.09805	10.199	.99523	24
37	.09787	.09834	10.168	.99520	23
38	.09816	.09864	10.138	.99517	22
39	.09845	.09893	10.108	.99514	21
40	.09874	.09923	10.078	.99511	20
41	.09903	.09952	10.048	.99508	19
42	.09932	.09981	10.019	.99506	18
43	.09961	.10011	9.9893	.99503	17
44	.09990	.10040	9.9601	.99500	16
45	.10019	.10069	9.9310	.99497	15
46	.10048	.10099	9.9021	.99494	14
47	.10077	.10128	9.8734	.99491	13
48	.10106	.10158	9.8448	.99488	12
49	.10135	.10187	9.8164	.99485	11
50	.10164	.10216	9.7882	.99482	10
51	.10192	.10246	9.7601	.99479	9
52	.10221	.10275	9.7322	.99476	8
53	.10250	.10305	9.7044	.99473	7
54	.10279	.10334	9.6768	.99470	6
55	.10308	.10363	9.6493	.99467	5
56	.10337	.10393	9.6220	.99464	4
57	.10366	.10422	9.5949	.99461	3
58	.10395	.10452	9.5679	.99458	2
59	.10424	.10481	9.5411	.99455	1
60	.10453	.10510	9.5144	.99452	0
′	Cos	Cot	Tan	Sin	′

95° (275°) — **(264°) 84°**

6° (186°) (353°) **173°**

′	Sin	Tan	Cot	Cos	′
0	.10453	.10510	9.5144	.99452	**60**
1	.10482	.10540	9.4878	.99449	59
2	.10511	.10569	9.4614	.99446	58
3	.10540	.10599	9.4352	.99443	57
4	.10569	.10628	9.4090	.99440	56
5	.10597	.10657	9.3831	.99437	**55**
6	.10626	.10687	9.3572	.99434	54
7	.10655	.10716	9.3315	.99431	53
8	.10684	.10746	9.3060	.99428	52
9	.10713	.10775	9.2806	.99424	51
10	.10742	.10805	9.2553	.99421	**50**
11	.10771	.10834	9.2302	.99418	49
12	.10800	.10863	9.2052	.99415	48
13	.10829	.10893	9.1803	.99412	47
14	.10858	.10922	9.1555	.99409	46
15	.10887	.10952	9.1309	.99406	**45**
16	.10916	.10981	9.1065	.99402	44
17	.10945	.11011	9.0821	.99399	43
18	.10973	.11040	9.0579	.99396	42
19	.11002	.11070	9.0338	.99393	41
20	.11031	.11099	9.0098	.99390	**40**
21	.11060	.11128	8.9860	.99386	39
22	.11089	.11158	8.9623	.99383	38
23	.11118	.11187	8.9387	.99380	37
24	.11147	.11217	8.9152	.99377	36
25	.11176	.11246	8.8919	.99374	**35**
26	.11205	.11276	8.8686	.99370	34
27	.11234	.11305	8.8455	.99367	33
28	.11263	.11335	8.8225	.99364	32
29	.11291	.11364	8.7996	.99360	31
30	.11320	.11394	8.7769	.99357	**30**
31	.11349	.11423	8.7542	.99354	29
32	.11378	.11452	8.7317	.99351	28
33	.11407	.11482	8.7093	.99347	27
34	.11436	.11511	8.6870	.99344	26
35	.11465	.11541	8.6648	.99341	**25**
36	.11494	.11570	8.6427	.99337	24
37	.11523	.11600	8.6208	.99334	23
38	.11552	.11629	8.5989	.99331	22
39	.11580	.11659	8.5772	.99327	21
40	.11609	.11688	8.5555	.99324	**20**
41	.11638	.11718	8.5340	.99320	19
42	.11667	.11747	8.5126	.99317	18
43	.11696	.11777	8.4913	.99314	17
44	.11725	.11806	8.4701	.99310	16
45	.11754	.11836	8.4490	.99307	**15**
46	.11783	.11865	8.4280	.99303	14
47	.11812	.11895	8.4071	.99300	13
48	.11840	.11924	8.3863	.99297	12
49	.11869	.11954	8.3656	.99293	11
50	.11898	.11983	8.3450	.99290	**10**
51	.11927	.12013	8.3245	.99286	9
52	.11956	.12042	8.3041	.99283	8
53	.11985	.12072	8.2838	.99279	7
54	.12014	.12101	8.2636	.99276	6
55	.12043	.12131	8.2434	.99272	**5**
56	.12071	.12160	8.2234	.99269	4
57	.12100	.12190	8.2035	.99265	3
58	.12129	.12219	8.1837	.99262	2
59	.12158	.12249	8.1640	.99258	1
60	.12187	.12278	8.1443	.99255	**0**
′	Cos	Cot	Tan	Sin	′

96° (276°) (263°) **83°**

7° (187°) (352°) **172°**

′	Sin	Tan	Cot	Cos	′
0	.12187	.12278	8.1443	.99255	**60**
1	.12216	.12308	8.1248	.99251	59
2	.12245	.12338	8.1054	.99248	58
3	.12274	.12367	8.0860	.99244	57
4	.12302	.12397	8.0667	.99240	56
5	.12331	.12426	8.0476	.99237	**55**
6	.12360	.12456	8.0285	.99233	54
7	.12389	.12485	8.0095	.99230	53
8	.12418	.12515	7.9906	.99226	52
9	.12447	.12544	7.9718	.99222	51
10	.12476	.12574	7.9530	.99219	**50**
11	.12504	.12603	7.9344	.99215	49
12	.12533	.12633	7.9158	.99211	48
13	.12562	.12662	7.8973	.99208	47
14	.12591	.12692	7.8789	.99204	46
15	.12620	.12722	7.8606	.99200	**45**
16	.12649	.12751	7.8424	.99197	44
17	.12678	.12781	7.8243	.99193	43
18	.12706	.12810	7.8062	.99189	42
19	.12735	.12840	7.7882	.99186	41
20	.12764	.12869	7.7704	.99182	**40**
21	.12793	.12899	7.7525	.99178	39
22	.12822	.12929	7.7348	.99175	38
23	.12851	.12958	7.7171	.99171	37
24	.12880	.12988	7.6996	.99167	36
25	.12908	.13017	7.6821	.99163	**35**
26	.12937	.13047	7.6647	.99160	34
27	.12966	.13076	7.6473	.99156	33
28	.12995	.13106	7.6301	.99152	32
29	.13024	.13136	7.6129	.99148	31
30	.13053	.13165	7.5958	.99144	**30**
31	.13081	.13195	7.5787	.99141	29
32	.13110	.13224	7.5618	.99137	28
33	.13139	.13254	7.5449	.99133	27
34	.13168	.13284	7.5281	.99129	26
35	.13197	.13313	7.5113	.99125	**25**
36	.13226	.13343	7.4947	.99122	24
37	.13254	.13372	7.4781	.99118	23
38	.13283	.13402	7.4615	.99114	22
39	.13312	.13432	7.4451	.99110	21
40	.13341	.13461	7.4287	.99106	**20**
41	.13370	.13491	7.4124	.99102	19
42	.13399	.13521	7.3962	.99098	18
43	.13427	.13550	7.3800	.99094	17
44	.13456	.13580	7.3639	.99091	16
45	.13485	.13609	7.3479	.99087	**15**
46	.13514	.13639	7.3319	.99083	14
47	.13543	.13669	7.3160	.99079	13
48	.13572	.13698	7.3002	.99075	12
49	.13600	.13728	7.2844	.99071	11
50	.13629	.13758	7.2687	.99067	**10**
51	.13658	.13787	7.2531	.99063	9
52	.13687	.13817	7.2375	.99059	8
53	.13716	.13846	7.2220	.99055	7
54	.13744	.13876	7.2066	.99051	6
55	.13773	.13906	7.1912	.99047	**5**
56	.13802	.13935	7.1759	.99043	4
57	.13831	.13965	7.1607	.99039	3
58	.13860	.13995	7.1455	.99035	2
59	.13889	.14024	7.1304	.99031	1
60	.13917	.14054	7.1154	.99027	**0**
′	Cos	Cot	Tan	Sin	′

97° (277°) (262°) **82°**

NATURAL TRIGONOMETRIC FUNCTIONS

8° (188°) (351°) **171°**

′	Sin	Tan	Cot	Cos	′
0	.13917	.14054	7.1154	.99027	**60**
1	.13946	.14084	7.1004	.99023	59
2	.13975	.14113	7.0855	.99019	58
3	.14004	.14143	7.0706	.99015	57
4	.14033	.14173	7.0558	.99011	56
5	.14061	.14202	7.0410	.99006	**55**
6	.14090	.14232	7.0264	.99002	54
7	.14119	.14262	7.0117	.98998	53
8	.14148	.14291	6.9972	.98994	52
9	.14177	.14321	6.9827	.98990	51
10	.14205	.14351	6.9682	.98986	**50**
11	.14234	.14381	6.9538	.98982	49
12	.14263	.14410	6.9395	.98978	48
13	.14292	.14440	6.9252	.98973	47
14	.14320	.14470	6.9110	.98969	46
15	.14349	.14499	6.8969	.98965	**45**
16	.14378	.14529	6.8828	.98961	44
17	.14407	.14559	6.8687	.98957	43
18	14436	.14588	6.8548	.98953	42
19	.14464	.14618	6.8408	.98948	41
20	.14493	.14648	6.8269	.98944	**40**
21	.14522	.14678	6.8131	.98940	39
22	.14551	.14707	6.7994	.98936	38
23	.14580	.14737	6.7856	.98931	37
24	.14608	.14767	6.7720	.98927	36
25	.14637	.14796	6.7584	.98923	**35**
26	.14666	.14826	6.7448	.98919	34
27	.14695	.14856	6.7313	.98914	33
28	.14723	.14886	6.7179	.98910	32
29	.14752	.14915	6.7045	.98906	31
30	.14781	.14945	6.6912	.98902	**30**
31	.14810	.14975	6.6779	.98897	29
32	.14838	.15005	6.6646	.98893	28
33	.14867	.15034	6.6514	.98889	27
34	.14896	.15064	6.6383	.98884	26
35	.14925	.15094	6.6252	.98880	**25**
36	.14954	.15124	6.6122	.98876	24
37	.14982	.15153	6.5992	.98871	23
38	.15011	.15183	6.5863	.98867	22
39	.15040	.15213	6.5734	.98863	21
40	.15069	.15243	6.5606	.98858	**20**
41	.15097	.15272	6.5478	.98854	19
42	.15126	.15302	6.5350	.98849	18
43	.15155	.15332	6.5223	.98845	17
44	.15184	.15362	6.5097	.98841	16
45	.15212	.15391	6.4971	.98836	**15**
46	.15241	.15421	6.4846	.98832	14
47	.15270	.15451	6.4721	.98827	13
48	.15299	.15481	6.4596	.98823	12
49	.15327	.15511	6.4472	.98818	11
50	.15356	.15540	6.4348	.98814	**10**
51	.15385	.15570	6.4225	.98809	9
52	.15414	.15600	6.4103	.98805	8
53	.15442	.15630	6.3980	.98800	7
54	.15471	.15660	6.3859	.98796	6
55	.15500	.15689	6.3737	.98791	**5**
56	.15529	.15719	6.3617	.98787	4
57	.15557	.15749	6.3496	.98782	3
58	.15586	.15779	6.3376	.98778	2
59	.15615	.15809	6.3257	.98773	1
60	.15643	.15838	6.3138	.98769	**0**
′	Cos	Cot	Tan	Sin	′

98° (278°) (261°) **81°**

9° (189°) (350°) **170°**

′	Sin	Tan	Cot	Cos	′
0	.15643	.15838	6.3138	.98769	**60**
1	.15672	.15868	6.3019	.98764	59
2	.15701	.15898	6.2901	.98760	58
3	.15730	.15928	6.2783	.98755	57
4	.15758	.15958	6.2666	.98751	56
5	.15787	.15988	6.2549	.98746	**55**
6	.15816	.16017	6.2432	.98741	54
7	.15845	.16047	6.2316	.98737	53
8	.15873	.16077	6.2200	.98732	52
9	.15902	.16107	6.2085	.98728	51
10	.15931	.16137	6.1970	.98723	**50**
11	.15959	.16167	6.1856	.98718	49
12	.15988	.16196	6.1742	.98714	48
13	.16017	.16226	6.1628	.98709	47
14	.16046	.16256	6.1515	.98704	46
15	.16074	.16286	6.1402	.98700	**45**
16	.16103	.16316	6.1290	.98695	44
17	.16132	.16346	6.1178	.98690	43
18	.16160	.16376	6.1066	.98686	42
19	.16189	.16405	6.0955	.98681	41
20	.16218	.16435	6.0844	.98676	**40**
21	.16246	.16465	6.0734	.98671	39
22	.16275	.16495	6.0624	.98667	38
23	.16304	.16525	6.0514	.98662	37
24	.16333	.16555	6.0405	.98657	36
25	.16361	.16585	6.0296	.98652	**35**
26	.16390	.16615	6.0188	.98648	34
27	.16419	.16645	6.0080	.98643	33
28	.16447	.16674	5.9972	.98638	32
29	.16476	.16704	5.9865	.98633	31
30	.16505	.16734	5.9758	.98629	**30**
31	.16533	.16764	5.9651	.98624	29
32	.16562	.16794	5.9545	.98619	28
33	.16591	.16824	5.9439	.98614	27
34	.16620	.16854	5.9333	.98609	26
35	.16648	.16884	5.9228	.98604	**25**
36	.16677	.16914	5.9124	.98600	24
37	.16706	.16944	5.9019	.98595	23
38	.16734	.16974	5.8915	.98590	22
39	.16763	.17004	5.8811	.98585	21
40	.16792	.17033	5.8708	.98580	**20**
41	.16820	.17063	5.8605	.98575	19
42	.16849	.17093	5.8502	.98570	18
43	.16878	.17123	5.8400	.98565	17
44	.16906	.17153	5.8298	.98561	16
45	.16935	.17183	5.8197	.98556	**15**
46	.16964	.17213	5.8095	.98551	14
47	.16992	.17243	5.7994	.98546	13
48	.17021	.17273	5.7894	.98541	12
49	.17050	.17303	5.7794	.98536	11
50	.17078	.17333	5.7694	.98531	**10**
51	.17107	.17363	5.7594	.98526	9
52	.17136	.17393	5.7495	.98521	8
53	.17164	.17423	5.7396	.98516	7
54	.17193	.17453	5.7297	.98511	6
55	.17222	.17483	5.7199	.98506	**5**
56	.17250	.17513	5.7101	.98501	4
57	.17279	.17543	5.7004	.98496	3
58	.17308	.17573	5.6906	.98491	2
59	.17336	.17603	5.6809	.98486	1
60	.17365	.17633	5.6713	.98481	**0**
′	Cos	Cot	Tan	Sin	′

99° (279°) (260°) **80°**

NATURAL TRIGONOMETRIC FUNCTIONS

10° (190°) (349°) **169°**

′	Sin	Tan	Cot	Cos	′
0	.17365	.17633	5.6713	.98481	**60**
1	.17393	.17663	5.6617	.98476	59
2	.17422	.17693	5.6521	.98471	58
3	.17451	.17723	5.6425	.98466	57
4	.17479	.17753	5.6329	.98461	56
5	.17508	.17783	5.6234	.98455	**55**
6	.17537	.17813	5.6140	.98450	54
7	.17565	.17843	5.6045	.98445	53
8	.17594	.17873	5.5951	.98440	52
9	.17623	.17903	5.5857	.98435	51
10	.17651	.17933	5.5764	.98430	**50**
11	.17680	.17963	5.5671	.98425	49
12	.17708	.17993	5.5578	.98420	48
13	.17737	.18023	5.5485	.98414	47
14	.17766	.18053	5.5393	.98409	46
15	.17794	.18083	5.5301	.98404	**45**
16	.17823	.18113	5.5209	.98399	44
17	.17852	.18143	5.5118	.98394	43
18	.17880	.18173	5.5026	.98389	42
19	.17909	.18203	5.4936	.98383	41
20	.17937	.18233	5.4845	.98378	**40**
21	.17966	.18263	5.4755	.98373	39
22	.17995	.18293	5.4665	.98368	38
23	.18023	.18323	5.4575	.98362	37
24	.18052	.18353	5.4486	.98357	36
25	.18081	.18384	5.4397	.98352	**35**
26	.18109	.18414	5.4308	.98347	34
27	.18138	.18444	5.4219	.98341	33
28	.18166	.18474	5.4131	.98336	32
29	.18195	.18504	5.4043	.98331	31
30	.18224	.18534	5.3955	.98325	**30**
31	.18252	.18564	5.3868	.98320	29
32	.18281	.18594	5.3781	.98315	28
33	.18309	.18624	5.3694	.98310	27
34	.18338	.18654	5.3607	.98304	26
35	.18367	.18684	5.3521	.98299	**25**
36	.18395	.18714	5.3435	.98294	24
37	.18424	.18745	5.3349	.98288	23
38	.18452	.18775	5.3263	.98283	22
39	.18481	.18805	5.3178	.98277	21
40	.18509	.18835	5.3093	.98272	**20**
41	.18538	.18865	5.3008	.98267	19
42	.18567	.18895	5.2924	.98261	18
43	.18595	.18925	5.2839	.98256	17
44	.18624	.18955	5.2755	.98250	16
45	.18652	.18986	5.2672	.98245	**15**
46	.18681	.19016	5.2588	.98240	14
47	.18710	.19046	5.2505	.98234	13
48	.18738	.19076	5.2422	.98229	12
49	.18767	.19106	5.2339	.98223	11
50	.18795	.19136	5.2257	.98218	**10**
51	.18824	.19166	5.2174	.98212	9
52	.18852	.19197	5.2092	.98207	8
53	.18881	.19227	5.2011	.98201	7
54	.18910	.19257	5.1929	.98196	6
55	.18938	.19287	5.1848	.98190	**5**
56	.18967	.13317	5.1767	.98185	4
57	.18995	.19347	5.1686	.98179	3
58	.19024	.19378	5.1606	.98174	2
59	.19052	.19408	5.1526	.98168	1
60	.19081	.19438	5.1446	.98163	**0**
′	Cos	Cot	Tan	Sin	′

100° (280°) (259°) **79°**

11° (191°) (348°) **168°**

′	Sin	Tan	Cot	Cos	′
0	.19081	.19438	5.1446	.98163	**60**
1	.19109	.19468	5.1366	.98157	59
2	.19138	.19498	5.1286	.98152	58
3	.19167	.19529	5.1207	.98146	57
4	.19195	.19559	5.1128	.98140	56
5	.19224	.19589	5.1049	.98135	**55**
6	.19252	.19619	5.0970	.98129	54
7	.19281	.19649	5.0892	.98124	53
8	.19309	.19680	5.0814	.98118	52
9	.19338	.19710	5.0736	.98112	51
10	.19366	.19740	5.0658	.98107	**50**
11	.19395	.19770	5.0581	.98101	49
12	.19423	.19801	5.0504	.98096	48
13	.19452	.19831	5.0427	.98090	47
14	.19481	.19861	5.0350	.98084	46
15	.19509	.19891	5.0273	.98079	**45**
16	.19538	.19921	5.0197	.98073	44
17	.19566	.19952	5.0121	.98067	43
18	.19595	.19982	5.0045	.98061	42
19	.19623	.20012	4.9969	.98056	41
20	.19652	.20042	4.9894	.98050	**40**
21	.19680	.20073	4.9819	.98044	39
22	.19709	.20103	4.9744	.98039	38
23	.19737	.20133	4.9669	.98033	37
24	.19766	.20164	4.9594	.98027	36
25	.19794	.20194	4.9520	.98021	**35**
26	.19823	.20224	4.9446	.98016	34
27	.19851	.20254	4.9372	.98010	33
28	.19880	.20285	4.9298	.98004	32
29	.19908	.20315	4.9225	.97998	31
30	.19937	.20345	4.9152	.97992	**30**
31	.19965	.20376	4.9078	.97987	29
32	.19994	.20406	4.9006	.97981	28
33	.20022	.20436	4.8933	.97975	27
34	.20051	.20466	4.8860	.97969	26
35	.20079	.20497	4.8788	.97963	**25**
36	.20108	.20527	4.8716	.97958	24
37	.20136	.20557	4.8644	.97952	23
38	.20165	.20588	4.8573	.97946	22
39	.20193	.20618	4.8501	.97940	21
40	.20222	.20648	4.8430	.97934	**20**
41	.20250	.20679	4.8359	.97928	19
42	.20279	.20709	4.8288	.97922	18
43	.20307	.20739	4.8218	.97916	17
44	.20336	.20770	4.8147	.97910	16
45	.20364	.20800	4.8077	.97905	**15**
46	.20393	.20830	4.8007	.97899	14
47	.20421	.20861	4.7937	.97893	13
48	.20450	.20891	4.7867	.97887	12
49	.20478	.20921	4.7798	.97881	11
50	.20507	.20952	4.7729	.97875	**10**
51	.20535	.20982	4.7659	.97869	9
52	.20563	.21013	4.7591	.97863	8
53	.20592	.21043	4.7522	.97857	7
54	.20620	.21073	4.7453	.97851	6
55	.20649	.21104	4.7385	.97845	**5**
56	.20677	.21134	4.7317	.97839	4
57	.20706	.21164	4.7249	.97833	3
58	.20734	.21195	4.7181	.97827	2
59	.20763	.21225	4.7114	.97821	1
60	.20791	.21256	4.7046	.97815	**0**
′	Cos	Cot	Tan	Sin	′

101° (281°) (258°) **78°**

NATURAL TRIGONOMETRIC FUNCTIONS

12° (192°) (347°) **167°**

′	Sin	Tan	Cot	Cos	′
0	.20791	.21256	4.7046	.97815	**60**
1	.20820	.21286	4.6979	.97809	59
2	.20848	.21316	4.6912	.97803	58
3	.20877	.21347	4.6845	.97797	57
4	.20905	.21377	4.6779	.97791	56
5	.20933	.21408	4.6712	.97784	**55**
6	.20962	.21438	4.6646	.97778	54
7	.20990	.21469	4.6580	.97772	53
8	.21019	.21499	4.6514	.97766	52
9	.21047	.21529	4.6448	.97760	51
10	.21076	.21560	4.6382	.97754	**50**
11	.21104	.21590	4.6317	.97748	49
12	.21132	.21621	4.6252	.97742	48
13	.21161	.21651	4.6187	.97735	47
14	.21189	.21682	4.6122	.97729	46
15	.21218	.21712	4.6057	.97723	**45**
16	.21246	.21743	4.5993	.97717	44
17	.21275	.21773	4.5928	.97711	43
18	.21303	.21804	4.5864	.97705	42
19	.21331	.21834	4.5800	.97698	41
20	.21360	.21864	4.5736	.97692	**40**
21	.21388	.21895	4.5673	.97686	39
22	.21417	.21925	4.5609	.97680	38
23	.21445	.21956	4.5546	.97673	37
24	.21474	.21986	4.5483	.97667	36
25	.21502	.22017	4.5420	.97661	**35**
26	.21530	.22047	4.5357	.97655	34
27	.21559	.22078	4.5294	.97648	33
28	.21587	.22108	4.5232	.97642	32
29	.21616	.22139	4.5169	.97636	31
30	.21644	.22169	4.5107	.97630	**30**
31	.21672	.22200	4.5045	.97623	29
32	.21701	.22231	4.4983	.97617	28
33	.21729	.22261	4.4922	.97611	27
34	.21758	.22292	4.4860	.97604	26
35	.21786	.22322	4.4799	.97598	**25**
36	.21814	.22353	4.4737	.97592	24
37	.21843	.22383	4.4676	.97585	23
38	.21871	.22414	4.4615	.97579	22
39	.21899	.22444	4.4555	.97573	21
40	.21928	.22475	4.4494	.97566	**20**
41	.21956	.22505	4.4434	.97560	19
42	.21985	.22536	4.4373	.97553	18
43	.22013	.22567	4.4313	.97547	17
44	.22041	.22597	4.4253	.97541	16
45	.22070	.22628	4.4194	.97534	**15**
46	.22098	.22658	4.4134	.97528	14
47	.22126	.22689	4.4075	.97521	13
48	.22155	.22719	4.4015	.97515	12
49	.22183	.22750	4.3956	.97508	11
50	.22212	.22781	4.3897	.97502	**10**
51	.22240	.22811	4.3838	.97496	9
52	.22268	.22842	4.3779	.97489	8
53	.22297	.22872	4.3721	.97483	7
54	.22325	.22903	4.3662	.97476	6
55	.22353	.22934	4.3604	.97470	**5**
56	.22382	.22964	4.3546	.97463	4
57	.22410	.22995	4.3488	.97457	3
58	.22438	.23026	4.3430	.97450	2
59	.22467	.23056	4.3372	.97444	1
60	.22495	.23087	4.3315	.97437	**0**
′	Cos	Cot	Tan	Sin	′

102° (282°) (257°) **77°**

13° (193°) (346°) **166°**

′	Sin	Tan	Cot	Cos	′
0	.22495	.23087	4.3315	.97437	**60**
1	.22523	.23117	4.3257	.97430	59
2	.22552	.23148	4.3200	.97424	58
3	.22580	.23179	4.3143	.97417	57
4	.22608	.23209	4.3086	.97411	56
5	.22637	.23240	4.3029	.97404	**55**
6	22665	.23271	4.2972	.97398	54
7	.22693	.23301	4.2916	.97391	53
8	.22722	.23332	4.2859	.97384	52
9	.22750	.23363	4.2803	.97378	51
10	.22778	.23393	4.2747	.97371	**50**
11	.22807	.23424	4.2691	.97365	49
12	.22835	.23455	4.2635	.97358	48
13	.22863	.23485	4.2580	.97351	47
14	.22892	.23516	4.2524	.97345	46
15	.22920	.23547	4.2468	.97338	**45**
16	.22948	.23578	4.2413	.97331	44
17	.22977	.23608	4.2358	.97325	43
18	.23005	.23639	4.2303	.97318	42
19	.23033	.23670	4.2248	.97311	41
20	.23062	.23700	4.2193	.97304	**40**
21	.23090	.23731	4.2139	.97298	39
22	.23118	.23762	4.2084	.97291	38
23	.23146	.23793	4.2030	.97284	37
24	.23175	.23823	4.1976	.97278	36
25	.23203	.23854	4.1922	.97271	**35**
26	.23231	.23885	4.1868	.97264	34
27	.23260	.23916	4.1814	.97257	33
28	.23288	.23946	4.1760	.97251	32
29	.23316	.23977	4.1706	.97244	31
30	.23345	.24008	4.1653	.97237	**30**
31	.23373	.24039	4.1600	.97230	29
32	.23401	.24069	4.1547	.97223	28
33	.23429	.24100	4.1493	.97217	27
34	.23458	.24131	4.1441	.97210	26
35	.23486	.24162	4.1388	.97203	**25**
36	.23514	.24193	4.1335	.97196	24
37	.23542	.24223	4.1282	.97189	23
38	.23571	.24254	4.1230	.97182	22
39	.23599	.24285	4.1178	.97176	21
40	.23627	.24316	4.1126	.97169	**20**
41	.23656	.24347	4.1074	.97162	19
42	.23684	.24377	4.1022	.97155	18
43	.23712	.24408	4.0970	.97148	17
44	.23740	.24439	4.0918	.97141	16
45	.23769	.24470	4.0867	.97134	**15**
46	.23797	.24501	4.0815	.97127	14
47	.23825	.24532	4.0764	.97120	13
48	.23853	.24562	4.0713	.97113	12
49	.23882	.24593	4.0662	.97106	11
50	.23910	.24624	4.0611	.97100	**10**
51	.23938	.24655	4.0560	.97093	9
52	.23966	.24686	4.0509	.97086	8
53	.23995	.24717	4.0459	.97079	7
54	.24023	.24747	4.0408	.97072	6
55	.24051	.24778	4.0358	.97065	**5**
56	.24079	.24809	4.0308	.97058	4
57	.24108	.24840	4.0257	.97051	3
58	.24136	.24871	4.0207	.97044	2
59	.24164	.24902	4.0158	.97037	1
60	.24192	.24933	4.0108	.97030	**0**
′	Cos	Cot	Tan	Sin	′

103° (283°) (256°) **76°**

14° (194°) (345°) **165°**

′	Sin	Tan	Cot	Cos	′
0	.24192	.24933	4.0108	.97030	**60**
1	.24220	.24964	4.0058	.97023	59
2	.24249	.24995	4.0009	.97015	58
3	.24277	.25026	3.9959	.97008	57
4	.24305	.25056	3.9910	.97001	56
5	.24333	.25087	3.9861	.96994	**55**
6	.24362	.25118	3.9812	.96987	54
7	.24390	.25149	3.9763	.96980	53
8	.24418	.25180	3.9714	.96973	52
9	.24446	.25211	3.9665	.96966	51
10	.24474	.25242	3.9617	.96959	**50**
11	.24503	.25273	3.9568	.96952	49
12	.24531	.25304	3.9520	.96945	48
13	.24559	.25335	3.9471	.96937	47
14	.24587	.25366	3.9423	.96930	46
15	.24615	.25397	3.9375	.96923	**45**
16	.24644	.25428	3.9327	.96916	44
17	.24672	.25459	3.9279	.96909	43
18	.24700	.25490	3.9232	.96902	42
19	.24728	.25521	3.9184	.96894	41
20	.24756	.25552	3.9136	.96887	**40**
21	.24784	.25583	3.9089	.96880	39
22	.24813	.25614	3.9042	.96873	38
23	.24841	.25645	3.8995	.96866	37
24	.24869	.25676	3.8947	.96858	36
25	.24897	.25707	3.8900	.96851	**35**
26	.24925	.25738	3.8854	.96844	34
27	.24954	.25769	3.8807	.96837	33
28	.24982	.25800	3.8760	.96829	32
29	.25010	.25831	3.8714	.96822	31
30	.25038	.25862	3.8667	.96815	**30**
31	.25066	.25893	3.8621	.96807	29
32	.25094	.25924	3.8575	.96800	28
33	.25122	.25955	3.8528	.96793	27
34	.25151	.25986	3.8482	.96786	26
35	.25179	.26017	3.8436	.96778	**25**
36	.25207	.26048	3.8391	.96771	24
37	.25235	.26079	3.8345	.96764	23
38	.25263	.26110	3.8299	.96756	22
39	.25291	.26141	3.8254	.96749	21
40	.25320	.26172	3.8208	.96742	**20**
41	.25348	.26203	3.8163	.96734	19
42	.25376	.26235	3.8118	.96727	18
43	.25404	.26266	3.8073	.96719	17
44	.25432	.26297	3.8028	.96712	16
45	.25460	.26328	3.7983	.96705	**15**
46	.25488	.26359	3.7938	.96697	14
47	.25516	.26390	3.7893	.96690	13
48	.25545	.26421	3.7848	.96682	12
49	.25573	.26452	3.7804	.96675	11
50	.25601	.26483	3.7760	.96667	**10**
51	.25629	.26515	3.7715	.96660	9
52	.25657	.26546	3.7671	.96653	8
53	.25685	.26577	3.7627	.96645	7
54	.25713	.26608	3.7583	.96638	6
55	.25741	.26639	3.7539	.96630	**5**
56	.25769	.26670	3.7495	.96623	4
57	.25798	.26701	3.7451	.96615	3
58	.25826	.26733	3.7408	.96608	2
59	.25854	.26764	3.7364	.96600	1
60	.25882	.26795	3.7321	.96593	**0**
′	Cos	Cot	Tan	Sin	′

104° (284°) (255°) **75°**

15° (195°) (344°) **164°**

′	Sin	Tan	Cot	Cos	′
0	.25882	.26795	3.7321	.96593	**60**
1	.25910	.26826	3.7277	.96585	59
2	.25938	.26857	3.7234	.96578	58
3	.25966	.26888	3.7191	.96570	57
4	.25994	.26920	3.7148	.96562	56
5	.26022	.26951	3.7105	.96555	**55**
6	.26050	.26982	3.7062	.96547	54
7	.26079	.27013	3.7019	.96540	53
8	.26107	.27044	3.6976	.96532	52
9	.26135	.27076	3.6933	.96524	51
10	.26163	.27107	3.6891	.96517	**50**
11	.26191	.27138	3.6848	.96509	49
12	.26219	.27169	3.6806	.96502	48
13	.26247	.27201	3.6764	.96494	47
14	.26275	.27232	3.6722	.96486	46
15	.26303	.27263	3.6680	.96479	**45**
16	.26331	.27294	3.6638	.96471	44
17	.26359	.27326	3.6596	.96463	43
18	.26387	.27357	3.6554	.96456	42
19	.26415	.27388	3.6512	.96448	41
20	.26443	.27419	3.6470	.96440	**40**
21	.26471	.27451	3.6429	.96433	39
22	.26500	.27482	3.6387	.96425	38
23	.26528	.27513	3.6346	.96417	37
24	.26556	.27545	3.6305	.96410	36
25	.26584	.27576	3.6264	.96402	**35**
26	.26612	.27607	3.6222	.96394	34
27	.26640	.27638	3.6181	.96386	33
28	.26668	.27670	3.6140	.96379	32
29	.26696	.27701	3.6100	.96371	31
30	.26724	.27732	3.6059	.96363	**30**
31	.26752	.27764	3.6018	.96355	29
32	.26780	.27795	3.5978	.96347	28
33	.26808	.27826	3.5937	.96340	27
34	.26836	.27858	3.5897	.96332	26
35	.26864	.27889	3.5856	.96324	**25**
36	.26892	.27921	3.5816	.96316	24
37	.26920	.27952	3.5776	.96308	23
38	.26948	.27983	3.5736	.96301	22
39	.26976	.28015	3.5696	.96293	21
40	.27004	.28046	3.5656	.96285	**20**
41	.27032	.28077	3.5616	.96277	19
42	.27060	.28109	3.5576	.96269	18
43	.27088	.28140	3.5536	.96261	17
44	.27116	.28172	3.5497	.96253	16
45	.27144	.28203	3.5457	.96246	**15**
46	.27172	.28234	3.5418	.96238	14
47	.27200	.28266	3.5379	.96230	13
48	.27228	.28297	3.5339	.96222	12
49	.27256	.28329	3.5300	.96214	11
50	.27284	.28360	3.5261	.96206	**10**
51	.27312	.28391	3.5222	.96198	9
52	.27340	.28423	3.5183	.96190	8
53	.27368	.28454	3.5144	.96182	7
54	.27396	.28486	3.5105	.96174	6
55	.27424	.28517	3.5067	.96166	**5**
56	.27452	.28549	3.5028	.96158	4
57	.27480	.28580	3.4989	.96150	3
58	.27508	.28612	3.4951	.96142	2
59	.27536	.28643	3.4912	.96134	1
60	.27564	.28675	3.4874	.96126	**0**
′	Cos	Cot	Tan	Sin	′

105° (285°) (254°) **74°**

NATURAL TRIGONOMETRIC FUNCTIONS

16° (196°) (343°) **163°**

′	Sin	Tan	Cot	Cos	′
0	.27564	.28675	3.4874	.96126	**60**
1	.27592	.28706	3.4836	.96118	59
2	.27620	.28738	3.4798	.96110	58
3	.27648	.28769	3.4760	.96102	57
4	.27676	.28801	3.4722	.96094	56
5	.27704	.28832	3.4684	.96086	**55**
6	.27731	.28864	3.4646	.96078	54
7	.27759	.28895	3.4608	.96070	53
8	.27787	.28927	3.4570	.96062	52
9	.27815	.28958	3.4533	.96054	51
10	.27843	.28990	3.4495	.96046	**50**
11	.27871	.29021	3.4458	.96037	49
12	.27899	.29053	3.4420	.96029	48
13	.27927	.29084	3.4383	.96021	47
14	.27955	.29116	3.4346	.96013	46
15	.27983	.29147	3.4308	.96005	**45**
16	.28011	.29179	3.4271	.95997	44
17	.28039	.29210	3.4234	.95989	43
18	.28067	.29242	3.4197	.95981	42
19	.28095	.29274	3.4160	.95972	41
20	.28123	.29305	3.4124	.95964	**40**
21	.28150	.29337	3.4087	.95956	39
22	.28178	.29368	3.4050	.95948	38
23	.28206	.29400	3.4014	.95940	37
24	.28234	.29432	3.3977	.95931	36
25	.28262	.29463	3.3941	.95923	**35**
26	.28290	.29495	3.3904	.95915	34
27	.28318	.29526	3.3868	.95907	33
28	.28346	.29558	3.3832	.95898	32
29	.28374	.29590	3.3796	.95890	31
30	.28402	.29621	3.3759	.95882	**30**
31	.28429	.29653	3.3723	.95874	29
32	.28457	.29685	3.3687	.95865	28
33	.28485	.29716	3.3652	.95857	27
34	.28513	.29748	3.3616	.95849	26
35	.28541	.29780	3.3580	.95841	**25**
36	.28569	.29811	3.3544	.95832	24
37	.28597	.29843	3.3509	.95824	23
38	.28625	.29875	3.3473	.95816	22
39	.28652	.29906	3.3438	.95807	21
40	.28680	.29938	3.3402	.95799	**20**
41	.28708	.29970	3.3367	.95791	19
42	.28736	.30001	3.3332	.95782	18
43	.28764	.30033	3.3297	.95774	17
44	.28792	.30065	3.3261	.95766	16
45	.28820	.30097	3.3226	.95757	**15**
46	.28847	.30128	3.3191	.95749	14
47	.28875	.30160	3.3156	.95740	13
48	.28903	.30192	3.3122	.95732	12
49	.28931	.30224	3.3087	.95724	11
50	.28959	.30255	3.3052	.95715	**10**
51	.28987	.30287	3.3017	.95707	9
52	.29015	.30319	3.2983	.95698	8
53	.29042	.30351	3.2948	.95690	7
54	.29070	.30382	3.2914	.95681	6
55	.29098	.30414	3.2879	.95673	**5**
56	.29126	.30446	3.2845	.95664	4
57	.29154	.30478	3.2811	.95656	3
58	.29182	.30509	3.2777	.95647	2
59	.29209	.30541	3.2743	.95639	1
60	.29237	.30573	3.2709	.95630	**0**
′	Cos	Cot	Tan	Sin	′

106° (286°) (253°) **73°**

17° (197°) (342°) **162°**

′	Sin	Tan	Cot	Cos	′
0	.29237	.30573	3.2709	.95630	**60**
1	.29265	.30605	3.2675	.95622	59
2	.29293	.30637	3.2641	.95613	58
3	.29321	.30669	3.2607	.95605	57
4	.29348	.30700	3.2573	.95596	56
5	.29376	.30732	3.2539	.95588	**55**
6	.29404	.30764	3.2506	.95579	54
7	.29432	.30796	3.2472	.95571	53
8	.29460	.30828	3.2438	.95562	52
9	.29487	.30860	3.2405	.95554	51
10	.29515	.30891	3.2371	.95545	**50**
11	.29543	.30923	3.2338	.95536	49
12	.29571	.30955	3.2305	.95528	48
13	.29599	.30987	3.2272	.95519	47
14	.29626	.31019	3.2238	.95511	46
15	.29654	.31051	3.2205	.95502	**45**
16	.29682	.31083	3.2172	.95493	44
17	.29710	.31115	3.2139	.95485	43
18	.29737	.31147	3.2106	.95476	42
19	.29765	.31178	3.2073	.95467	41
20	.29793	.31210	3.2041	.95459	**40**
21	.29821	.31242	3.2008	.95450	39
22	.29849	.31274	3.1975	.95441	38
23	.29876	.31306	3.1943	.95433	37
24	.29904	.31338	3.1910	.95424	36
25	.29932	.31370	3.1878	.95415	**35**
26	.29960	.31402	3.1845	.95407	34
27	.29987	.31434	3.1813	.95398	33
28	.30015	.31466	3.1780	.95389	32
29	.30043	.31498	3.1748	.95380	31
30	.30071	.31530	3.1716	.95372	**30**
31	.30098	.31562	3.1684	.95363	29
32	.30126	.31594	3.1652	.95354	28
33	.30154	.31626	3.1620	.95345	27
34	.30182	.31658	3.1588	.95337	26
35	.30209	.31690	3.1556	.95328	**25**
36	.30237	.31722	3.1524	.95319	24
37	.30265	.31754	3.1492	.95310	23
38	.30292	.31786	3.1460	.95301	22
39	.30320	.31818	3.1429	.95293	21
40	.30348	.31850	3.1397	.95284	**20**
41	.30376	.31882	3.1366	.95275	19
42	.30403	.31914	3.1334	.95266	18
43	.30431	.31946	3.1303	.95257	17
44	.30459	.31978	3.1271	.95248	16
45	.30486	.32010	3.1240	.95240	**15**
46	.30514	.32042	3.1209	.95231	14
47	.30542	.32074	3.1178	.95222	13
48	.30570	.32106	3.1146	.95213	12
49	.30597	.32139	3.1115	.95204	11
50	.30625	.32171	3.1084	.95195	**10**
51	.30653	.32203	3.1053	.95186	9
52	.30680	.32235	3.1022	.95177	8
53	.30708	.32267	3.0991	.95168	7
54	.30736	.32299	3.0961	.95159	6
55	.30763	.32331	3.0930	.95150	**5**
56	.30791	.32363	3.0899	.95142	4
57	.30819	.32396	3.0868	.95133	3
58	.30846	.32428	3.0838	.95124	2
59	.30874	.32460	3.0807	.95115	1
60	.30902	.32492	3.0777	.95106	**0**
′	Cos	Cot	Tan	Sin	′

107° (287°) (252°) **72°**

NATURAL TRIGONOMETRIC FUNCTIONS

18° (198°) (341°) **161°**

′	Sin	Tan	Cot	Cos	′
0	.30902	.32492	3.0777	.95106	**60**
1	.30929	.32524	3.0746	.95097	59
2	.30957	.32556	3.0716	.95088	58
3	.30985	.32588	3.0686	.95079	57
4	.31012	.32621	3.0655	.95070	56
5	.31040	.32653	3.0625	.95061	**55**
6	.31068	.32685	3.0595	.95052	54
7	.31095	.32717	3.0565	.95043	53
8	.31123	.32749	3.0535	.95033	52
9	.31151	.32782	3.0505	.95024	51
10	.31178	.32814	3.0475	.95015	**50**
11	.31206	.32846	3.0445	.95006	49
12	.31233	.32878	3.0415	.94997	48
13	.31261	.32911	3.0385	.94988	47
14	.31289	.32943	3.0356	.94979	46
15	.31316	.32975	3.0326	.94970	**45**
16	.31344	.33007	3.0296	.94961	44
17	.31372	.33040	3.0267	.94952	43
18	.31399	.33072	3.0237	.94943	42
19	.31427	.33104	3.0208	.94933	41
20	.31454	.33136	3.0178	.94924	**40**
21	.31482	.33169	3.0149	.94915	39
22	.31510	.33201	3.0120	.94906	38
23	.31537	.33233	3.0090	.94897	37
24	.31565	.33266	3.0061	.94888	36
25	.31593	.33298	3.0032	.94878	**35**
26	.31620	.33330	3.0003	.94869	34
27	.31648	.33363	2.9974	.94860	33
28	.31675	.33395	2.9945	.94851	32
29	.31703	.33427	2.9916	.94842	31
30	.31730	.33460	2.9887	.94832	**30**
31	.31758	.33492	2.9858	.94823	29
32	.31786	.33524	2.9829	.94814	28
33	.31813	.33557	2.9800	.94805	27
34	.31841	.33589	2.9772	.94795	26
35	.31868	.33621	2.9743	.94786	**25**
36	.31896	.33654	2.9714	.94777	24
37	.31923	.33686	2.9686	.94768	23
38	.31951	.33718	2.9657	.94758	22
39	.31979	.33751	2.9629	.94749	21
40	.32006	.33783	2.9600	.94740	**20**
41	.32034	.33816	2.9572	.94730	19
42	.32061	.33848	2.9544	.94721	18
43	.32089	.33881	2.9515	.94712	17
44	.32116	.33913	2.9487	.94702	16
45	.32144	.33945	2.9459	.94693	**15**
46	.32171	.33978	2.9431	.94684	14
47	.32199	.34010	2.9403	.94674	13
48	.32227	.34043	2.9375	.94665	12
49	.32254	.34075	2.9347	.94656	11
50	.32282	.34108	2.9319	.94646	**10**
51	.32309	.34140	2.9291	.94637	9
52	.32337	.34173	2.9263	.94627	8
53	.32364	.34205	2.9235	.94618	7
54	.32392	.34238	2.9208	.94609	6
55	.32419	.34270	2.9180	.94599	**5**
56	.32447	.34303	2.9152	.94590	4
57	.32474	.34335	2.9125	.94580	3
58	.32502	.34368	2.9097	.94571	2
59	.32529	.34400	2.9070	.94561	1
60	.32557	.34433	2.9042	.94552	**0**
′	Cos	Cot	Tan	Sin	′

108° (288°) (251°) **71°**

19° (199°) (340°) **160°**

′	Sin	Tan	Cot	Cos	′
0	.32557	.34433	2.9042	.94552	**60**
1	.32584	.34465	2.9015	.94542	59
2	.32612	.34498	2.8987	.94533	58
3	.32639	.34530	2.8960	.94523	57
4	.32667	.34563	2.8933	.94514	56
5	.32694	.34596	2.8905	.94504	**55**
6	.32722	.34628	2.8878	.94495	54
7	.32749	.34661	2.8851	.94485	53
8	.32777	.34693	2.8824	.94476	52
9	.32804	.34726	2.8797	.94466	51
10	.32832	.34758	2.8770	.94457	**50**
11	.32859	.34791	2.8743	.94447	49
12	.32887	.34824	2.8716	.94438	48
13	.32914	.34856	2.8689	.94428	47
14	.32942	.34889	2.8662	.94418	46
15	.32969	.34922	2.8636	.94409	**45**
16	.32997	.34954	2.8609	.94399	44
17	.33024	.34987	2.8582	.94390	43
18	.33051	.35020	2.8556	.94380	42
19	.33079	.35052	2.8529	.94370	41
20	.33106	.35085	2.8502	.94361	**40**
21	.33134	.35118	2.8476	.94351	39
22	.33161	.35150	2.8449	.94342	38
23	.33189	.35183	2.8423	.94332	37
24	.33216	.35216	2.8397	.94322	36
25	.33244	.35248	2.8370	.94313	**35**
26	.33271	.35281	2.8344	.94303	34
27	.33298	.35314	2.8318	.94293	33
28	.33326	.35346	2.8291	.94284	32
29	.33353	.35379	2.8265	.94274	31
30	.33381	.35412	2.8239	.94264	**30**
31	.33408	.35445	2.8213	.94254	29
32	.33436	.35477	2.8187	.94245	28
33	.33463	.35510	2.8161	.94235	27
34	.33490	.35543	2.8135	.94225	26
35	.33518	.35576	2.8109	.94215	**25**
36	.33545	.35608	2.8083	.94206	24
37	.33573	.35641	2.8057	.94196	23
38	.33600	.35674	2.8032	.94186	22
39	.33627	.35707	2.8006	.94176	21
40	.33655	.35740	2.7980	.94167	**20**
41	.33682	.35772	2.7955	.94157	19
42	.33710	.35805	2.7929	.94147	18
43	.33737	.35838	2.7903	.94137	17
44	.33764	.35871	2.7878	.94127	16
45	.33792	.35904	2.7852	.94118	**15**
46	.33819	.35937	2.7827	.94108	14
47	.33846	.35969	2.7801	.94098	13
48	.33874	.36002	2.7776	.94088	12
49	.33901	.36035	2.7751	.94078	11
50	.33929	.36068	2.7725	.94068	**10**
51	.33956	.36101	2.7700	.94058	9
52	.33983	.36134	2.7675	.94049	8
53	.34011	.36167	2.7650	.94039	7
54	.34038	.36199	2.7625	.94029	6
55	.34065	.36232	2.7600	.94019	**5**
56	.34093	.36265	2.7575	.94009	4
57	.34120	.36298	2.7550	.93999	3
58	.34147	.36331	2.7525	.93989	2
59	.34175	.36364	2.7500	.93979	1
60	.34202	.36397	2.7475	.93969	**0**
′	Cos	Cot	Tan	Sin	′

109° (289°) (250°) **70°**

20° (200°) (339°) **159°**

′	Sin	Tan	Cot	Cos	′
0	.34202	.36397	2.7475	.93969	**60**
1	.34229	.36430	2.7450	.93959	59
2	.34257	.36463	2.7425	.93949	58
3	.34284	.36496	2.7400	.93939	57
4	.34311	.36529	2.7376	.93929	56
5	.34339	.36562	2.7351	.93919	**55**
6	.34366	.36595	2.7326	.93909	54
7	.34393	.36628	2.7302	.93899	53
8	.34421	.36661	2.7277	.93889	52
9	.34448	.36694	2.7253	.93879	51
10	.34475	.36727	2.7228	.93869	**50**
11	.34503	.36760	2.7204	.93859	49
12	.34530	.36793	2.7179	.93849	48
13	.34557	.36826	2.7155	.93839	47
14	.34584	.36859	2.7130	.93829	46
15	.34612	.36892	2.7106	.93819	**45**
16	.34639	.36925	2.7082	.93809	44
17	.34666	.36958	2.7058	.93799	43
18	.34694	.36991	2.7034	.93789	42
19	.34721	.37024	2.7009	.93779	41
20	.34748	.37057	2.6985	.93769	**40**
21	.34775	.37090	2.6961	.93759	39
22	.34803	.37123	2.6937	.93748	38
23	.34830	.37157	2.6913	.93738	37
24	.34857	.37190	2.6889	.93728	36
25	.34884	.37223	2.6865	.93718	**35**
26	.34912	.37256	2.6841	.93708	34
27	.34939	.37289	2.6818	.93698	33
28	.34966	.37322	2.6794	.93688	32
29	.34993	.37355	2.6770	.93677	31
30	.35021	.37388	2.6746	.93667	**30**
31	.35048	.37422	2.6723	.93657	29
32	.35075	.37455	2.6699	.93647	28
33	.35102	.37488	2.6675	.93637	27
34	.35130	.37521	2.6652	.93626	26
35	.35157	.37554	2.6628	.93616	**25**
36	.35184	.37588	2.6605	.93606	24
37	.35211	.37621	2.6581	.93596	23
38	.35239	.37654	2.6558	.93585	22
39	.35266	.37687	2.6534	.93575	21
40	.35293	.37720	2.6511	.93565	**20**
41	.35320	.37754	2.6488	.93555	19
42	.35347	.37787	2.6464	.93544	18
43	.35375	.37820	2.6441	.93534	17
44	.35402	.37853	2.6418	.93524	16
45	.35429	.37887	2.6395	.93514	**15**
46	.35456	.37920	2.6371	.93503	14
47	.35484	.37953	2.6348	.93493	13
48	.35511	.37986	2.6325	.93483	12
49	.35538	.38020	2.6302	.93472	11
50	.35565	.38053	2.6279	.93462	**10**
51	.35592	.38086	2.6256	.93452	9
52	.35619	.38120	2.6233	.93441	8
53	.35647	.38153	2.6210	.93431	7
54	.35674	.38186	2.6187	.93420	6
55	.35701	.38220	2.6165	.93410	**5**
56	.35728	.38253	2.6142	.93400	4
57	.35755	.38286	2.6119	.93389	3
58	.35782	.38320	2.6096	.93379	2
59	.35810	.38353	2.6074	.93368	1
60	.35837	.38386	2.6051	.93358	**0**
′	Cos	Cot	Tan	Sin	′

110° (290°) (249°) **69°**

21° (201°) (338°) **158°**

′	Sin	Tan	Cot	Cos	′
0	.35837	.38386	2.6051	.93358	**60**
1	.35864	.38420	2.6028	.93348	59
2	.35891	.38453	2.6006	.93337	58
3	.35918	.38487	2.5983	.93327	57
4	.35945	.38520	2.5961	.93316	56
5	.35973	.38553	2.5938	.93306	**55**
6	.36000	.38587	2.5916	.93295	54
7	.36027	.38620	2.5893	.93285	53
8	.36054	.38654	2.5871	.93274	52
9	.36081	.38687	2.5848	.93264	51
10	.36108	.38721	2.5826	.93253	**50**
11	.36135	.38754	2.5804	.93243	49
12	.36162	.38787	2.5782	.93232	48
13	.36190	.38821	2.5759	.93222	47
14	.36217	.38854	2.5737	.93211	46
15	.36244	.38888	2.5715	.93201	**45**
16	.36271	.38921	2.5693	.93190	44
17	.36298	.38955	2.5671	.93180	43
18	.36325	.38988	2.5649	.93169	42
19	.36352	.39022	2.5627	.93159	41
20	.36379	.39055	2.5605	.93148	**40**
21	.36406	.39089	2.5583	.93137	39
22	.36434	.39122	2.5561	.93127	38
23	.36461	.39156	2.5539	.93116	37
24	.36488	.39190	2.5517	.93106	36
25	.36515	.39223	2.5495	.93095	**35**
26	.36542	.39257	2.5473	.93084	34
27	.36569	.39290	2.5452	.93074	33
28	.36596	.39324	2.5430	.93063	32
29	.36623	.39357	2.5408	.93052	31
30	.36650	.39391	2.5386	.93042	**30**
31	.36677	.39425	2.5365	.93031	29
32	.36704	.39458	2.5343	.93020	28
33	.36731	.39492	2.5322	.93010	27
34	.36758	.39526	2.5300	.92999	26
35	.36785	.39559	2.5279	.92988	**25**
36	.36812	.39593	2.5257	.92978	24
37	.36839	.39626	2.5236	.92967	23
38	.36867	.39660	2.5214	.92956	22
39	.36894	.39694	2.5193	.92945	21
40	.36921	.39727	2.5172	.92935	**20**
41	.36948	.39761	2.5150	.92924	19
42	.36975	.39795	2.5129	.92913	18
43	.37002	.39829	2.5108	.92902	17
44	.37029	.39862	2.5086	.92892	16
45	.37056	.39896	2.5065	.92881	**15**
46	.37083	.39930	2.5044	.92870	14
47	.37110	.39963	2.5023	.92859	13
48	.37137	.39997	2.5002	.92849	12
49	.37164	.40031	2.4981	.92838	11
50	.37191	.40065	2.4960	.92827	**10**
51	.37218	.40098	2.4939	.92816	9
52	.37245	.40132	2.4918	.92805	8
53	.37272	.40166	2.4897	.92794	7
54	.37299	.40200	2.4876	.92784	6
55	.37326	.40234	2.4855	.92773	**5**
56	.37353	.40267	2.4834	.92762	4
57	.37380	.40301	2.4813	.92751	3
58	.37407	.40335	2.4792	.92740	2
59	.37434	.40369	2.4772	.92729	1
60	.37461	.40403	2.4751	.92718	**0**
′	Cos	Cot	Tan	Sin	′

111° (291°) (248°) **68°**

NATURAL TRIGONOMETRIC FUNCTIONS

22° (202°) (337°) **157°**

′	Sin	Tan	Cot	Cos	′
0	.37461	.40403	2.4751	.92718	**60**
1	.37488	.40436	2.4730	.92707	59
2	.37515	.40470	2.4709	.92697	58
3	.37542	.40504	2.4689	.92686	57
4	.37569	.40538	2.4668	.92675	56
5	.37595	.40572	2.4648	.92664	**55**
6	.37622	.40606	2.4627	.92653	54
7	.37649	.40640	2.4606	.92642	53
8	.37676	.40674	2.4586	.92631	52
9	.37703	.40707	2.4566	.92620	51
10	.37730	.40741	2.4545	.92609	**50**
11	.37757	.40775	2.4525	.92598	49
12	.37784	.40809	2.4504	.92587	48
13	.37811	.40843	2.4484	.92576	47
14	.37838	.40877	2.4464	.92565	46
15	.37865	.40911	2.4443	.92554	**45**
16	.37892	.40945	2.4423	.92543	44
17	.37919	.40979	2.4403	.92532	43
18	.37946	.41013	2.4383	.92521	42
19	.37973	.41047	2.4362	.92510	41
20	.37999	.41081	2.4342	.92499	**40**
21	.38026	.41115	2.4322	.92488	39
22	.38053	.41149	2.4302	.92477	38
23	.38080	.41183	2.4282	.92466	37
24	.38107	.41217	2.4262	.92455	36
25	.38134	.41251	2.4242	.92444	**35**
26	.38161	.41285	2.4222	.92432	34
27	.38188	.41319	2.4202	.92421	33
28	.38215	.41353	2.4182	.92410	32
29	.38241	.41387	2.4162	.92399	31
30	.38268	.41421	2.4142	.92388	**30**
31	.38295	.41455	2.4122	.92377	29
32	.38322	.41490	2.4102	.92366	28
33	.38349	.41524	2.4083	.92355	27
34	.38376	.41558	2.4063	.92343	26
35	.38403	.41592	2.4043	.92332	**25**
36	.38430	.41626	2.4023	.92321	24
37	.38456	.41660	2.4004	.92310	23
38	.38483	.41694	2.3984	.92299	22
39	.38510	.41728	2.3964	.92287	21
40	.38537	.41763	2.3945	.92276	**20**
41	.38564	.41797	2.3925	.92265	19
42	.38591	.41831	2.3906	.92254	18
43	.38617	.41865	2.3886	.92243	17
44	.38644	.41899	2.3867	.92231	16
45	.38671	.41933	2.3847	.92220	**15**
46	.38698	.41968	2.3828	.92209	14
47	.38725	.42002	2.3808	.92198	13
48	.38752	.42036	2.3789	.92186	12
49	.38778	.42070	2.3770	.92175	11
50	.38805	.42105	2.3750	.92164	**10**
51	.38832	.42139	2.3731	.92152	9
52	.38859	.42173	2.3712	.92141	8
53	.38886	.42207	2.3693	.92130	7
54	.38912	.42242	2.3673	.92119	6
55	.38939	.42276	2.3654	.92107	**5**
56	.38966	.42310	2.3635	.92096	4
57	.38993	.42345	2.3616	.92085	3
58	.39020	.42379	2.3597	.92073	2
59	.39046	.42413	2.3578	.92062	1
60	.39073	.42447	2.3559	.92050	**0**
′	Cos	Cot	Tan	Sin	′

112° (292°) (247°) **67°**

23° (203°) (336°) **156°**

′	Sin	Tan	Cot	Cos	′
0	.39073	.42447	2.3559	.92050	**60**
1	.39100	.42482	2.3539	.92039	59
2	.39127	.42516	2.3520	.92028	58
3	.39153	.42551	2.3501	.92016	57
4	.39180	.42585	2.3483	.92005	56
5	.39207	.42619	2.3464	.91994	**55**
6	.39234	.42654	2.3445	.91982	54
7	.39260	.42688	2.3426	.91971	53
8	.39287	.42722	2.3407	.91959	52
9	.39314	.42757	2.3388	.91948	51
10	.39341	.42791	2.3369	.91936	**50**
11	.39367	.42826	2.3351	.91925	49
12	.39394	.42860	2.3332	.91914	48
13	.39421	.42894	2.3313	.91902	47
14	.39448	.42929	2.3294	.91891	46
15	.39474	.42963	2.3276	.91879	**45**
16	.39501	.42998	2.3257	.91868	44
17	.39528	.43032	2.3238	.91856	43
18	.39555	.43067	2.3220	.91845	42
19	.39581	.43101	2.3201	.91833	41
20	.39608	.43136	2.3183	.91822	**40**
21	.39635	.43170	2.3164	.91810	39
22	.39661	.43205	2.3146	.91799	38
23	.39688	.43239	2.3127	.91787	37
24	.39715	.43274	2.3109	.91775	36
25	.39741	.43308	2.3090	.91764	**35**
26	.39768	.43343	2.3072	.91752	34
27	.39795	.43378	2.3053	.91741	33
28	.39822	.43412	2.3035	.91729	32
29	.39848	.43447	2.3017	.91718	31
30	.39875	.43481	2.2998	.91706	**30**
31	.39902	.43516	2.2980	.91694	29
32	.39928	.43550	2.2962	.91683	28
33	.39955	.43585	2.2944	.91671	27
34	.39982	.43620	2.2925	.91660	26
35	.40008	.43654	2.2907	.91648	**25**
36	.40035	.43689	2.2889	.91636	24
37	.40062	.43724	2.2871	.91625	23
38	.40088	.43758	2.2853	.91613	22
39	.40115	.43793	2.2835	.91601	21
40	.40141	.43828	2.2817	.91590	**20**
41	.40168	.43862	2.2799	.91578	19
42	.40195	.43897	2.2781	.91566	18
43	.40221	.43932	2.2763	.91555	17
44	.40248	.43966	2.2745	.91543	16
45	.40275	.44001	2.2727	.91531	**15**
46	.40301	.44036	2.2709	.91519	14
47	.40328	.44071	2.2691	.91508	13
48	.40355	.44105	2.2673	.91496	12
49	.40381	.44140	2.2655	.91484	11
50	.40408	.44175	2.2637	.91472	**10**
51	.40434	.44210	2.2620	.91461	9
52	.40461	.44244	2.2602	.91449	8
53	.40488	.44279	2.2584	.91437	7
54	.40514	.44314	2.2566	.91425	6
55	.40541	.44349	2.2549	.91414	**5**
56	.40567	.44384	2.2531	.91402	4
57	.40594	.44418	2.2513	.91390	3
58	.40621	.44453	2.2496	.91378	2
59	.40647	.44488	2.2478	.91366	1
60	.40674	.44523	2.2460	.91355	**0**
′	Cos	Cot	Tan	Sin	′

113° (293°) (246°) **66°**

NATURAL TRIGONOMETRIC FUNCTIONS

24° (204°) (335°) **155°**

′	Sin	Tan	Cot	Cos	′
0	.40674	.44523	2.2460	.91355	**60**
1	.40700	.44558	2.2443	.91343	59
2	.40727	.44593	2.2425	.91331	58
3	.40753	.44627	2.2408	.91319	57
4	.40780	.44662	2.2390	.91307	56
5	.40806	.44697	2.2373	.91295	**55**
6	.40833	.44732	2.2355	.91283	54
7	.40860	.44767	2.2338	.91272	53
8	.40886	.44802	2.2320	.91260	52
9	.40913	.44837	2.2303	.91248	51
10	.40939	.44872	2.2286	.91236	**50**
11	.40966	.44907	2.2268	.91224	49
12	.40992	.44942	2.2251	.91212	48
13	.41019	.44977	2.2234	.91200	47
14	.41045	.45012	2.2216	.91188	46
15	.41072	.45047	2.2199	.91176	**45**
16	.41098	.45082	2.2182	.91164	44
17	.41125	.45117	2.2165	.91152	43
18	.41151	.45152	2.2148	.91140	42
19	.41178	.45187	2.2130	.91128	41
20	.41204	.45222	2.2113	.91116	**40**
21	.41231	.45257	2.2096	.91104	39
22	.41257	.45292	2.2079	.91092	38
23	.41284	.45327	2.2062	.91080	37
24	.41310	.45362	2.2045	.91068	36
25	.41337	.45397	2.2028	.91056	**35**
26	.41363	.45432	2.2011	.91044	34
27	.41390	.45467	2.1994	.91032	33
28	.41416	.45502	2.1977	.91020	32
29	.41443	.45538	2.1960	.91008	31
30	.41469	.45573	2.1943	.90996	**30**
31	.41496	.45608	2.1926	.90984	29
32	.41522	.45643	2.1909	.90972	28
33	.41549	.45678	2.1892	.90960	27
34	.41575	.45713	2.1876	.90948	26
35	.41602	.45748	2.1859	.90936	**25**
36	.41628	.45784	2.1842	.90924	24
37	.41655	.45819	2.1825	.90911	23
38	.41681	.45854	2.1808	.90899	22
39	.41707	.45889	2.1792	.90887	21
40	.41734	.45924	2.1775	.90875	**20**
41	.41760	.45960	2.1758	.90863	19
42	.41787	.45995	2.1742	.90851	18
43	.41813	.46030	2.1725	.90839	17
44	.41840	.46065	2.1708	.90826	16
45	.41866	.46101	2.1692	.90814	**15**
46	.41892	.46136	2.1675	.90802	14
47	.41919	.46171	2.1659	.90790	13
48	.41945	.46206	2.1642	.90778	12
49	.41972	.46242	2.1625	.90766	11
50	.41998	.46277	2.1609	.90753	**10**
51	.42024	.46312	2.1592	.90741	9
52	.42051	.46348	2.1576	.90729	8
53	.42077	.46383	2.1560	.90717	7
54	.42104	.46418	2.1543	.90704	6
55	.42130	.46454	2.1527	.90692	**5**
56	.42156	.46489	2.1510	.90680	4
57	.42183	.46525	2.1494	.90668	3
58	.42209	.46560	2.1478	.90655	2
59	.42235	.46595	2.1461	.90643	1
60	.42262	.46631	2.1445	.90631	**0**
′	Cos	Cot	Tan	Sin	′

114° (294°) (245°) **65°**

25° (205°) (334°) **154°**

′	Sin	Tan	Cot	Cos	′
0	.42262	.46631	2.1445	.90631	**60**
1	.42288	.46666	2.1429	.90618	59
2	.42315	.46702	2.1413	.90606	58
3	.42341	.46737	2.1396	.90594	57
4	.42367	.46772	2.1380	.90582	56
5	.42394	.46808	2.1364	.90569	**55**
6	.42420	.46843	2.1348	.90557	54
7	.42446	.46879	2.1332	.90545	53
8	.42473	.46914	2.1315	.90532	52
9	.42499	.46950	2.1299	.90520	51
10	.42525	.46985	2.1283	.90507	**50**
11	.42552	.47021	2.1267	.90495	49
12	.42578	.47056	2.1251	.90483	48
13	.42604	.47092	2.1235	.90470	47
14	.42631	.47128	2.1219	.90458	46
15	.42657	.47163	2.1203	.90446	**45**
16	.42683	.47199	2.1187	.90433	44
17	.42709	.47234	2.1171	.90421	43
18	.42736	.47270	2.1155	.90408	42
19	.42762	.47305	2.1139	.90396	41
20	.42788	.47341	2.1123	.90383	**40**
21	.42815	.47377	2.1107	.90371	39
22	.42841	.47412	2.1092	.90358	38
23	.42867	.47448	2.1076	.90346	37
24	.42894	.47483	2.1060	.90334	36
25	.42920	.47519	2.1044	.90321	**35**
26	.42946	.47555	2.1028	.90309	34
27	.42972	.47590	2.1013	.90296	33
28	.42999	.47626	2.0997	.90284	32
29	.43025	.47662	2.0981	.90271	31
30	.43051	.47698	2.0965	.90259	**30**
31	.43077	.47733	2.0950	.90246	29
32	.43104	.47769	2.0934	.90233	28
33	.43130	.47805	2.0918	.90221	27
34	.43156	.47840	2.0903	.90208	26
35	.43182	.47876	2.0887	.90196	**25**
36	.43209	.47912	2.0872	.90183	24
37	.43235	.47948	2.0856	.90171	23
38	.43261	.47984	2.0840	.90158	22
39	.43287	.48019	2.0825	.90146	21
40	.43313	.48055	2.0809	.90133	**20**
41	.43340	.48091	2.0794	.90120	19
42	.43366	.48127	2.0778	.90108	18
43	.43392	.48163	2.0763	.90095	17
44	.43418	.48198	2.0748	.90082	16
45	.43445	.48234	2.0732	.90070	**15**
46	.43471	.48270	2.0717	.90057	14
47	.43497	.48306	2.0701	.90045	13
48	.43523	.48342	2.0686	.90032	12
49	.43549	.48378	2.0671	.90019	11
50	.43575	.48414	2.0655	.90007	**10**
51	.43602	.48450	2.0640	.89994	9
52	.43628	.48486	2.0625	.89981	8
53	.43654	.48521	2.0609	.89968	7
54	.43680	.48557	2.0594	.89956	6
55	.43706	.48593	2.0579	.89943	**5**
56	.43733	.48629	2.0564	.89930	4
57	.43759	.48665	2.0549	.89918	3
58	.43785	.48701	2.0533	.89905	2
59	.43811	.48737	2.0518	.89892	1
60	.43837	.48773	2.0503	.89879	**0**
′	Cos	Cot	Tan	Sin	′

115° (295°) (244°) **64°**

NATURAL TRIGONOMETRIC FUNCTIONS

26° (206°) (333°) **153°**

′	Sin	Tan	Cot	Cos	′
0	.43837	.48773	2.0503	.89879	**60**
1	.43863	.48809	2.0488	.89867	59
2	.43889	.48845	2.0473	.89854	58
3	.43916	.48881	2.0458	.89841	57
4	.43942	.48917	2.0443	.89828	56
5	.43968	.48953	2.0428	.89816	**55**
6	.43994	.48989	2.0413	.89803	54
7	.44020	.49026	2.0398	.89790	53
8	.44046	.49062	2.0383	.89777	52
9	.44072	.49098	2.0368	.89764	51
10	.44098	.49134	2.0353	.89752	**50**
11	.44124	.49170	2.0338	.89739	49
12	.44151	.49206	2.0323	.89726	48
13	.44177	.49242	2.0308	.89713	47
14	.44203	.49278	2.0293	.89700	46
15	.44229	.49315	2.0278	.89687	**45**
16	.44255	.49351	2.0263	.89674	44
17	.44281	.49387	2.0248	.89662	43
18	.44307	.49423	2.0233	.89649	42
19	.44333	.49459	2.0219	.89636	41
20	.44359	.49495	2.0204	.89623	**40**
21	.44385	.49532	2.0189	.89610	39
22	.44411	.49568	2.0174	.89597	38
23	.44437	.49604	2.0160	.89584	37
24	.44464	.49640	2.0145	.89571	36
25	.44490	.49677	2.0130	.89558	**35**
26	.44516	.49713	2.0115	.89545	34
27	.44542	.49749	2.0101	.89532	33
28	.44568	.49786	2.0086	.89519	32
29	.44594	.49822	2.0072	.89506	31
30	.44620	.49858	2.0057	.89493	**30**
31	.44646	.49894	2.0042	.89480	29
32	.44672	.49931	2.0028	.89467	28
33	.44698	.49967	2.0013	.89454	27
34	.44724	.50004	1.9999	.89441	26
35	.44750	.50040	1.9984	.89428	**25**
36	.44776	.50076	1.9970	.89415	24
37	.44802	.50113	1.9955	.89402	23
38	.44828	.50149	1.9941	.89389	22
39	.44854	.50185	1.9926	.89376	21
40	.44880	.50222	1.9912	.89363	**20**
41	.44906	.50258	1.9897	.89350	19
42	.44932	.50295	1.9883	.89337	18
43	.44958	.50331	1.9868	.89324	17
44	.44984	.50368	1.9854	.89311	16
45	.45010	.50404	1.9840	.89298	**15**
46	.45036	.50441	1.9825	.89285	14
47	.45062	.50477	1.9811	.89272	13
48	.45088	.50514	1.9797	.89259	12
49	.45114	.50550	1.9782	.89245	11
50	.45140	.50587	1.9768	.89232	**10**
51	.45166	.50623	1.9754	.89219	9
52	.45192	.50660	1.9740	.89206	8
53	.45218	.50696	1.9725	.89193	7
54	.45243	.50733	1.9711	.89180	6
55	.45269	.50769	1.9697	.89167	**5**
56	.45295	.50806	1.9683	.89153	4
57	.45321	.50843	1.9669	.89140	3
58	.45347	.50879	1.9654	.89127	2
59	.45373	.50916	1.9640	.89114	1
60	.45399	.50953	1.9626	.89101	**0**
′	Cos	Cot	Tan	Sin	′

116° (296°) (243°) **63°**

27° (207°) (332°) **152°**

′	Sin	Tan	Cot	Cos	′
0	.45399	.50953	1.9626	.89101	**60**
1	.45425	.50989	1.9612	.89087	59
2	.45451	.51026	1.9598	.89074	58
3	.45477	.51063	1.9584	.89061	57
4	.45503	.51099	1.9570	.89048	56
5	.45529	.51136	1.9556	.89035	**55**
6	.45554	.51173	1.9542	.89021	54
7	.45580	.51209	1.9528	.89008	53
8	.45606	.51246	1.9514	.88995	52
9	.45632	.51283	1.9500	.88981	51
10	.45658	.51319	1.9486	.88968	**50**
11	.45684	.51356	1.9472	.88955	49
12	.45710	.51393	1.9458	.88942	48
13	.45736	.51430	1.9444	.88928	47
14	.45762	.51467	1.9430	.88915	46
15	.45787	.51503	1.9416	.88902	**45**
16	.45813	.51540	1.9402	.88888	44
17	.45839	.51577	1.9388	.88875	43
18	.45865	.51614	1.9375	.88862	42
19	.45891	.51651	1.9361	.88848	41
20	.45917	.51688	1.9347	.88835	**40**
21	.45942	.51724	1.9333	.88822	39
22	.45968	.51761	1.9319	.88808	38
23	.45994	.51798	1.9306	.88795	37
24	.46020	.51835	1.9292	.88782	36
25	.46046	.51872	1.9278	.88768	**35**
26	.46072	.51909	1.9265	.88755	34
27	.46097	.51946	1.9251	.88741	33
28	.46123	.51983	1.9237	.88728	32
29	.46149	.52020	1.9223	.88715	31
30	.46175	.52057	1.9210	.88701	**30**
31	.46201	.52094	1.9196	.88688	29
32	.46226	.52131	1.9183	.88674	28
33	.46252	.52168	1.9169	.88661	27
34	.46278	.52205	1.9155	.88647	26
35	.46304	.52242	1.9142	.88634	**25**
36	.46330	.52279	1.9128	.88620	24
37	.46355	.52316	1.9115	.88607	23
38	.46381	.52353	1.9101	.88593	22
39	.46407	.52390	1.9088	.88580	21
40	.46433	.52427	1.9074	.88566	**20**
41	.46458	.52464	1.9061	.88553	19
42	.46484	.52501	1.9047	.88539	18
43	.46510	.52538	1.9034	.88526	17
44	.46536	.52575	1.9020	.88512	16
45	.46561	.52613	1.9007	.88499	**15**
46	.46587	.52650	1.8993	.88485	14
47	.46613	.52687	1.8980	.88472	13
48	.46639	.52724	1.8967	.88458	12
49	.46664	.52761	1.8953	.88445	11
50	.46690	.52798	1.8940	.88431	**10**
51	.46716	.52836	1.8927	.88417	9
52	.46742	.52873	1.8913	.88404	8
53	.46767	.52910	1.8900	.88390	7
54	.46793	.52947	1.8887	.88377	6
55	.46819	.52985	1.8873	.88363	**5**
56	.46844	.53022	1.8860	.88349	4
57	.46870	.53059	1.8847	.88336	3
58	.46896	.53096	1.8834	.88322	2
59	.46921	.53134	1.8820	.88308	1
60	.46947	.53171	1.8807	.88295	**0**
′	Cos	Cot	Tan	Sin	′

117° (297°) (242°) **62°**

NATURAL TRIGONOMETRIC FUNCTIONS

28° (208°) (331°) **151°**

′	Sin	Tan	Cot	Cos	′
0	.46947	.53171	1.8807	.88295	**60**
1	.46973	.53208	1.8794	.88281	59
2	.46999	.53246	1.8781	.88267	58
3	.47024	.53283	1.8768	.88254	57
4	.47050	.53320	1.8755	.88240	56
5	.47076	.53358	1.8741	.88226	**55**
6	.47101	.53395	1.8728	.88213	54
7	.47127	.53432	1.8715	.88199	53
8	.47153	.53470	1.8702	.88185	52
9	.47178	.53507	1.8689	.88172	51
10	.47204	.53545	1.8676	.88158	**50**
11	.47229	.53582	1.8663	.88144	49
12	.47255	.53620	1.8650	.88130	48
13	.47281	.53657	1.8637	.88117	47
14	.47306	.53694	1.8624	.88103	46
15	.47332	.53732	1.8611	.88089	**45**
16	.47358	.53769	1.8598	.88075	44
17	.47383	.53807	1.8585	.88062	43
18	.47409	.53844	1.8572	.88048	42
19	.47434	.53882	1.8559	.88034	41
20	.47460	.53920	1.8546	.88020	**40**
21	.47486	.53957	1.8533	.88006	39
22	.47511	.53995	1.8520	.87993	38
23	.47537	.54032	1.8507	.87979	37
24	.47562	.54070	1.8495	.87965	36
25	.47588	.54107	1.8482	.87951	**35**
26	.47614	.54145	1.8469	.87937	34
27	.47639	.54183	1.8456	.87923	33
28	.47665	.54220	1.8443	.87909	32
29	.47690	.54258	1.8430	.87896	31
30	.47716	.54296	1.8418	.87882	**30**
31	.47741	.54333	1.8405	.87868	29
32	.47767	.54371	1.8392	.87854	28
33	.47793	.54409	1.8379	.87840	27
34	.47818	.54446	1.8367	.87826	26
35	.47844	.54484	1.8354	.87812	**25**
36	.47869	.54522	1.8341	.87798	24
37	.47895	.54560	1.8329	.87784	23
38	.47920	.54597	1.8316	.87770	22
39	.47946	.54635	1.8303	.87756	21
40	.47971	.54673	1.8291	.87743	**20**
41	.47997	.54711	1.8278	.87729	19
42	.48022	.54748	1.8265	.87715	18
43	.48048	.54786	1.8253	.87701	17
44	.48073	.54824	1.8240	.87687	16
45	.48099	.54862	1.8228	.87673	**15**
46	.48124	.54900	1.8215	.87659	14
47	.48150	.54938	1.8202	.87645	13
48	.48175	.54975	1.8190	.87631	12
49	.48201	.55013	1.8177	.87617	11
50	.48226	.55051	1.8165	.87603	**10**
51	.48252	.55089	1.8152	.87589	9
52	.48277	.55127	1.8140	.87575	8
53	.48303	.55165	1.8127	.87561	7
54	.48328	.55203	1.8115	.87546	6
55	.48354	.55241	1.8103	.87532	**5**
56	.48379	.55279	1.8090	.87518	4
57	.48405	.55317	1.8078	.87504	3
58	.48430	.55355	1.8065	.87490	2
59	.48456	.55393	1.8053	.87476	1
60	.48481	.55431	1.8040	.87462	**0**
′	Cos	Cot	Tan	Sin	′

118° (298°) (241°) **61°**

29° (209°) (330°) **150°**

′	Sin	Tan	Cot	Cos	′
0	.48481	.55431	1.8040	.87462	**60**
1	.48506	.55469	1.8028	.87448	59
2	.48532	.55507	1.8016	.87434	58
3	.48557	.55545	1.8003	.87420	57
4	.48583	.55583	1.7991	.87406	56
5	.48608	.55621	1.7979	.87391	**55**
6	.48634	.55659	1.7966	.87377	54
7	.48659	.55697	1.7954	.87363	53
8	.48684	.55736	1.7942	.87349	52
9	.48710	.55774	1.7930	.87335	51
10	.48735	.55812	1.7917	.87321	**50**
11	.48761	.55850	1.7905	.87306	49
12	.48786	.55888	1.7893	.87292	48
13	.48811	.55926	1.7881	.87278	47
14	.48837	.55964	1.7868	.87264	46
15	.48862	.56003	1.7856	.87250	**45**
16	.48888	.56041	1.7844	.87235	44
17	.48913	.56079	1.7832	.87221	43
18	.48938	.56117	1.7820	.87207	42
19	.48964	.56156	1.7808	.87193	41
20	.48989	.56194	1.7796	.87178	**40**
21	.49014	.56232	1.7783	.87164	39
22	.49040	.56270	1.7771	.87150	38
23	.49065	.56309	1.7759	.87136	37
24	.49090	.56347	1.7747	.87121	36
25	.49116	.56385	1.7735	.87107	**35**
26	.49141	.56424	1.7723	.87093	34
27	.49166	.56462	1.7711	.87079	33
28	.49192	.56501	1.7699	.87064	32
29	.49217	.56539	1.7687	.87050	31
30	.49242	.56577	1.7675	.87036	**30**
31	.49268	.56616	1.7663	.87021	29
32	.49293	.56654	1.7651	.87007	28
33	.49318	.56693	1.7639	.86993	27
34	.49344	.56731	1.7627	.86978	26
35	.49369	.56769	1.7615	.86964	**25**
36	.49394	.56808	1.7603	.86949	24
37	.49419	.56846	1.7591	.86935	23
38	.49445	.56885	1.7579	.86921	22
39	.49470	.56923	1.7567	.86906	21
40	.49495	.56962	1.7556	.86892	**20**
41	.49521	.57000	1.7544	.86878	19
42	.49546	.57039	1.7532	.86863	18
43	.49571	.57078	1.7520	.86849	17
44	.49596	.57116	1.7508	.86834	16
45	.49622	.57155	1.7496	.86820	**15**
46	.49647	.57193	1.7485	.86805	14
47	.49672	.57232	1.7473	.86791	13
48	.49697	.57271	1.7461	.86777	12
49	.49723	.57309	1.7449	.86762	11
50	.49748	.57348	1.7437	.86748	**10**
51	.49773	.57386	1.7426	.86733	9
52	.49798	.57425	1.7414	.86719	8
53	.49824	.57464	1.7402	.86704	7
54	.49849	.57503	1.7391	.86690	6
55	.49874	.57541	1.7379	.86675	**5**
56	.49899	.57580	1.7367	.86661	4
57	.49924	.57619	1.7355	.86646	3
58	.49950	.57657	1.7344	.86632	2
59	.49975	.57696	1.7332	.86617	1
60	.50000	.57735	1.7321	.86603	**0**
′	Cos	Cot	Tan	Sin	′

119° (299°) (240°) **60°**

NATURAL TRIGONOMETRIC FUNCTIONS

30° (210°) (329°) **149°**

′	Sin	Tan	Cot	Cos	′
0	.50000	.57735	1.7321	.86603	**60**
1	.50025	.57774	1.7309	.86588	59
2	.50050	.57813	1.7297	.86573	58
3	.50076	.57851	1.7286	.86559	57
4	.50101	.57890	1.7274	.86544	56
5	.50126	.57929	1.7262	.86530	**55**
6	.50151	.57968	1.7251	.86515	54
7	.50176	.58007	1.7239	.86501	53
8	.50201	.58046	1.7228	.86486	52
9	.50227	.58085	1.7216	.86471	51
10	.50252	.58124	1.7205	.86457	**50**
11	.50277	.58162	1.7193	.86442	49
12	.50302	.58201	1.7182	.86427	48
13	.50327	.58240	1.7170	.86413	47
14	.50352	.58279	1.7159	.86398	46
15	.50377	.58318	1.7147	.86384	**45**
16	.50403	.58357	1.7136	.86369	44
17	.50428	.58396	1.7124	.86354	43
18	.50453	.58435	1.7113	.86340	42
19	.50478	.58474	1.7102	.86325	41
20	.50503	.58513	1.7090	.86310	**40**
21	.50528	.58552	1.7079	.86295	39
22	.50553	.58591	1.7067	.86281	38
23	.50578	.58631	1.7056	.86266	37
24	.50603	.58670	1.7045	.86251	36
25	.50628	.58709	1.7033	.86237	**35**
26	.50654	.58748	1.7022	.86222	34
27	.50679	.58787	1.7011	.86207	33
28	.50704	.58826	1.6999	.86192	32
29	.50729	.58865	1.6988	.86178	31
30	.50754	.58905	1.6977	.86163	**30**
31	.50779	.58944	1.6965	.86148	29
32	.50804	.58983	1.6954	.86133	28
33	.50829	.59022	1.6943	.86119	27
34	.50854	.59061	1.6932	.86104	26
35	.50879	.59101	1.6920	.86089	**25**
36	.50904	.59140	1.6909	.86074	24
37	.50929	.59179	1.6898	.86059	23
38	.50954	.59218	1.6887	.86045	22
39	.50979	.59258	1.6875	.86030	21
40	.51004	.59297	1.6864	.86015	**20**
41	.51029	.59336	1.6853	.86000	19
42	.51054	.59376	1.6842	.85985	18
43	.51079	.59415	1.6831	.85970	17
44	.51104	.59454	1.6820	.85956	16
45	.51129	.59494	1.6808	.85941	**15**
46	.51154	.59533	1.6797	.85926	14
47	.51179	.59573	1.6786	.85911	13
48	.51204	.59612	1.6775	.85896	12
49	.51229	.59651	1.6764	.85881	11
50	.51254	.59691	1.6753	.85866	**10**
51	.51279	.59730	1.6742	.85851	9
52	.51304	.59770	1.6731	.85836	8
53	.51329	.59809	1.6720	.85821	7
54	.51354	.59849	1.6709	.85806	6
55	.51379	.59888	1.6698	.85792	**5**
56	.51404	.59928	1.6687	.85777	4
57	.51429	.59967	1.6676	.85762	3
58	.51454	.60007	1.6665	.85747	2
59	.51479	.60046	1.6654	.85732	1
60	.51504	.60086	1.6643	.85717	**0**
′	Cos	Cot	Tan	Sin	′

120° (300°) (239°) **59°**

31° (211°) (328°) **148°**

′	Sin	Tan	Cot	Cos	′
0	.51504	.60086	1.6643	.85717	**60**
1	.51529	.60126	1.6632	.85702	59
2	.51554	.60165	1.6621	.85687	58
3	.51579	.60205	1.6610	.85672	57
4	.51604	.60245	1.6599	.85657	56
5	.51628	.60284	1.6588	.85642	**55**
6	.51653	.60324	1.6577	.85627	54
7	.51678	.60364	1.6566	.85612	53
8	.51703	.60403	1.6555	.85597	52
9	.51728	.60443	1.6545	.85582	51
10	.51753	.60483	1.6534	.85567	**50**
11	.51778	.60522	1.6523	.85551	49
12	.51803	.60562	1.6512	.85536	48
13	.51828	.60602	1.6501	.85521	47
14	.51852	.60642	1.6490	.85506	46
15	.51877	.60681	1.6479	.85491	**45**
16	.51902	.60721	1.6469	.85476	44
17	.51927	.60761	1.6458	.85461	43
18	.51952	.60801	1.6447	.85446	42
19	.51977	.60841	1.6436	.85431	41
20	.52002	.60881	1.6426	.85416	**40**
21	.52026	.60921	1.6415	.85401	39
22	.52051	.60960	1.6404	.85385	38
23	.52076	.61000	1.6393	.85370	37
24	.52101	.61040	1.6383	.85355	36
25	.52126	.61080	1.6372	.85340	**35**
26	.52151	.61120	1.6361	.85325	34
27	.52175	.61160	1.6351	.85310	33
28	.52200	.61200	1.6340	.85294	32
29	.52225	.61240	1.6329	.85279	31
30	.52250	.61280	1.6319	.85264	**30**
31	.52275	.61320	1.6308	.85249	29
32	.52299	.61360	1.6297	.85234	28
33	.52324	.61400	1.6287	.85218	27
34	.52349	.61440	1.6276	.85203	26
35	.52374	.61480	1.6265	.85188	**25**
36	.52399	.61520	1.6255	.85173	24
37	.52423	.61561	1.6244	.85157	23
38	.52448	.61601	1.6234	.85142	22
39	.52473	.61641	1.6223	.85127	21
40	.52498	.61681	1.6212	.85112	**20**
41	.52522	.61721	1.6202	.85096	19
42	.52547	.61761	1.6191	.85081	18
43	.52572	.61801	1.6181	.85066	17
44	.52597	.61842	1.6170	.85051	16
45	.52621	.61882	1.6160	.85035	**15**
46	.52646	.61922	1.6149	.85020	14
47	.52671	.61962	1.6139	.85005	13
48	.52696	.62003	1.6128	.84989	12
49	.52720	.62043	1.6118	.84974	11
50	.52745	.62083	1.6107	.84959	**10**
51	.52770	.62124	1.6097	.84943	9
52	.52794	.62164	1.6087	.84928	8
53	.52819	.62204	1.6076	.84913	7
54	.52844	.62245	1.6066	.84897	6
55	.52869	.62285	1.6055	.84882	**5**
56	.52893	.62325	1.6045	.84866	4
57	.52918	.62366	1.6034	.84851	3
58	.52943	.62406	1.6024	.84836	2
59	.52967	.62446	1.6014	.84820	1
60	.52992	.62487	1.6003	.84805	**0**
′	Cos	Cot	Tan	Sin	′

121° (301°) (238°) **58°**

NATURAL TRIGONOMETRIC FUNCTIONS

32° (212°) — (327°) **147°**

′	Sin	Tan	Cot	Cos	′
0	.52992	.62487	1.6003	.84805	**60**
1	.53017	.62527	1.5993	.84789	59
2	.53041	.62568	1.5983	.84774	58
3	.53066	.62608	1.5972	.84759	57
4	.53091	.62649	1.5962	.84743	56
5	.53115	.62689	1.5952	.84728	**55**
6	.53140	.62730	1.5941	.84712	54
7	.53164	.62770	1.5931	.84697	53
8	.53189	.62811	1.5921	.84681	52
9	.53214	.62852	1.5911	.84666	51
10	.53238	.62892	1.5900	.84650	**50**
11	.53263	.62933	1.5890	.84635	49
12	.53288	.62973	1.5880	.84619	48
13	.53312	.63014	1.5869	.84604	47
14	.53337	.63055	1.5859	.84588	46
15	.53361	.63095	1.5849	.84573	**45**
16	.53386	.63136	1.5839	.84557	44
17	.53411	.63177	1.5829	.84542	43
18	.53435	.63217	1.5818	.84526	42
19	.53460	.63258	1.5808	.84511	41
20	.53484	.63299	1.5798	.84495	**40**
21	.53509	.63340	1.5788	.84480	39
22	.53534	.63380	1.5778	.84464	38
23	.53558	.63421	1.5768	.84448	37
24	.53583	.63462	1.5757	.84433	36
25	.53607	.63503	1.5747	.84417	**35**
26	.53632	.63544	1.5737	.84402	34
27	.53656	.63584	1.5727	.84386	33
28	.53681	.63625	1.5717	.84370	32
29	.53705	.63666	1.5707	.84355	31
30	.53730	.63707	1.5697	.84339	**30**
31	.53754	.63748	1.5687	.84324	29
32	.53779	.63789	1.5677	.84308	28
33	.53804	.63830	1.5667	.84292	27
34	.53828	.63871	1.5657	.84277	26
35	.53853	.63912	1.5647	.84261	**25**
36	.53877	.63953	1.5637	.84245	24
37	.53902	.63994	1.5627	.84230	23
38	.53926	.64035	1.5617	.84214	22
39	.53951	.64076	1.5607	.84198	21
40	.53975	.64117	1.5597	.84182	**20**
41	.54000	.64158	1.5587	.84167	19
42	.54024	.64199	1.5577	.84151	18
43	.54049	.64240	1.5567	.84135	17
44	.54073	.64281	1.5557	.84120	16
45	.54097	.64322	1.5547	.84104	**15**
46	.54122	.64363	1.5537	.84088	14
47	.54146	.64404	1.5527	.84072	13
48	.54171	.64446	1.5517	.84057	12
49	.54195	.64487	1.5507	.84041	11
50	.54220	.64528	1.5497	.84025	**10**
51	.54244	.64569	1.5487	.84009	9
52	.54269	.64610	1.5477	.83994	8
53	.54293	.64652	1.5468	.83978	7
54	.54317	.64693	1.5458	.83962	6
55	.54342	.64734	1.5448	.83946	**5**
56	.54366	.64775	1.5438	.83930	4
57	.54391	.64817	1.5428	.83915	3
58	.54415	.64858	1.5418	.83899	2
59	.54440	.64899	1.5408	.83883	1
60	.54464	.64941	1.5399	.83867	**0**
′	Cos	Cot	Tan	Sin	′

122° (302°) — (237°) **57°**

33° (213°) — (326°) **146°**

′	Sin	Tan	Cot	Cos	′
0	.54464	.64941	1.5399	.83867	**60**
1	.54488	.64982	1.5389	.83851	59
2	.54513	.65024	1.5379	.83835	58
3	.54537	.65065	1.5369	.83819	57
4	.54561	.65106	1.5359	.83804	56
5	.54586	.65148	1.5350	.83788	**55**
6	.54610	.65189	1.5340	.83772	54
7	.54635	.65231	1.5330	.83756	53
8	.54659	.65272	1.5320	.83740	52
9	.54683	.65314	1.5311	.83724	51
10	.54708	.65355	1.5301	.83708	**50**
11	.54732	.65397	1.5291	.83692	49
12	.54756	.65438	1.5282	.83676	48
13	.54781	.65480	1.5272	.83660	47
14	.54805	.65521	1.5262	.83645	46
15	.54829	.65563	1.5253	.83629	**45**
16	.54854	.65604	1.5243	.83613	44
17	.54878	.65646	1.5233	.83597	43
18	.54902	.65688	1.5224	.83581	42
19	.54927	.65729	1.5214	.83565	41
20	.54951	.65771	1.5204	.83549	**40**
21	.54975	.65813	1.5195	.83533	39
22	.54999	.65854	1.5185	.83517	38
23	.55024	.65896	1.5175	.83501	37
24	.55048	.65938	1.5166	.83485	36
25	.55072	.65980	1.5156	.83469	**35**
26	.55097	.66021	1.5147	.83453	34
27	.55121	.66063	1.5137	.83437	33
28	.55145	.66105	1.5127	.83421	32
29	.55169	.66147	1.5118	.83405	31
30	.55194	.66189	1.5108	.83389	**30**
31	.55218	.66230	1.5099	.83373	29
32	.55242	.66272	1.5089	.83356	28
33	.55266	.66314	1.5080	.83340	27
34	.55291	.66356	1.5070	.83324	26
35	.55315	.66398	1.5061	.83308	**25**
36	.55339	.66440	1.5051	.83292	24
37	.55363	.66482	1.5042	.83276	23
38	.55388	.66524	1.5032	.83260	22
39	.55412	.66566	1.5023	.83244	21
40	.55436	.66608	1.5013	.83228	**20**
41	.55460	.66650	1.5004	.83212	19
42	.55484	.66692	1.4994	.83195	18
43	.55509	.66734	1.4985	.83179	17
44	.55533	.66776	1.4975	.83163	16
45	.55557	.66818	1.4966	.83147	**15**
46	.55581	.66860	1.4957	.83131	14
47	.55605	.66902	1.4947	.83115	13
48	.55630	.66944	1.4938	.83098	12
49	.55654	.66986	1.4928	.83082	11
50	.55678	.67028	1.4919	.83066	**10**
51	.55702	.67071	1.4910	.83050	9
52	.55726	.67113	1.4900	.83034	8
53	.55750	.67155	1.4891	.83017	7
54	.55775	.67197	1.4882	.83001	6
55	.55799	.67239	1.4872	.82985	**5**
56	.55823	.67282	1.4863	.82969	4
57	.55847	.67324	1.4854	.82953	3
58	.55871	.67366	1.4844	.82936	2
59	.55895	.67409	1.4835	.82920	1
60	.55919	.67451	1.4826	.82904	**0**
′	Cos	Cot	Tan	Sin	′

123° (303°) — (236°) **56°**

NATURAL TRIGONOMETRIC FUNCTIONS

34° (214°) (325°) **145°**

′	Sin	Tan	Cot	Cos	′
0	.55919	.67451	1.4826	.82904	**60**
1	.55943	.67493	1.4816	.82887	59
2	.55968	.67536	1.4807	.82871	58
3	.55992	.67578	1.4798	.82855	57
4	.56016	.67620	1.4788	.82839	56
5	.56040	.67663	1.4779	.82822	**55**
6	.56064	.67705	1.4770	.82806	54
7	.56088	.67748	1.4761	.82790	53
8	.56112	.67790	1.4751	.82773	52
9	.56136	.67832	1.4742	.82757	51
10	.56160	.67875	1.4733	.82741	**50**
11	.56184	.67917	1.4724	.82724	49
12	.56208	.67960	1.4715	.82708	48
13	.56232	.68002	1.4705	.82692	47
14	.56256	.68045	1.4696	.82675	46
15	.56280	.68088	1.4687	.82659	**45**
16	.56305	.68130	1.4678	.82643	44
17	.56329	.68173	1.4669	.82626	43
18	.56353	.68215	1.4659	.82610	42
19	.56377	.68258	1.4650	.82593	41
20	.56401	.68301	1.4641	.82577	**40**
21	.56425	.68343	1.4632	.82561	39
22	.56449	.68386	1.4623	.82544	38
23	.56473	.68429	1.4614	.82528	37
24	.56497	.68471	1.4605	.82511	36
25	.56521	.68514	1.4596	.82495	**35**
26	.56545	.68557	1.4586	.82478	34
27	.56569	.68600	1.4577	.82462	33
28	.56593	.68642	1.4568	.82446	32
29	.56617	.68685	1.4559	.82429	31
30	.56641	.68728	1.4550	.82413	**30**
31	.56665	.68771	1.4541	.82396	29
32	.56689	.68814	1.4532	.82380	28
33	.56713	.68857	1.4523	.82363	27
34	.56736	.68900	1.4514	.82347	26
35	.56760	.68942	1.4505	.82330	**25**
36	.56784	.68985	1.4496	.82314	24
37	.56808	.69028	1.4487	.82297	23
38	.56832	.69071	1.4478	.82281	22
39	.56856	.69114	1.4469	.82264	21
40	.56880	.69157	1.4460	.82248	**20**
41	.56904	.69200	1.4451	.82231	19
42	.56928	.69243	1.4442	.82214	18
43	.56952	.69286	1.4433	.82198	17
44	.56976	.69329	1.4424	.82181	16
45	.57000	.69372	1.4415	.82165	**15**
46	.57024	.69416	1.4406	.82148	14
47	.57047	.69459	1.4397	.82132	13
48	.57071	.69502	1.4388	.82115	12
49	.57095	.69545	1.4379	.82098	11
50	.57119	.69588	1.4370	.82082	**10**
51	.57143	.69631	1.4361	.82065	9
52	.57167	.69675	1.4352	.82048	8
53	.57191	.69718	1.4344	.82032	7
54	.57215	.69761	1.4335	.82015	6
55	.57238	.69804	1.4326	.81999	**5**
56	.57262	.69847	1.4317	.81982	4
57	.57286	.69891	1.4308	.81965	3
58	.57310	.69934	1.4299	.81949	2
59	.57334	.69977	1.4290	.81932	1
60	.57358	.70021	1.4281	.81915	**0**
′	Cos	Cot	Tan	Sin	′

124° (304°) (235°) **55°**

35° (215°) (324°) **144°**

′	Sin	Tan	Cot	Cos	′
0	.57358	.70021	1.4281	.81915	**60**
1	.57381	.70064	1.4273	.81899	59
2	.57405	.70107	1.4264	.81882	58
3	.57429	.70151	1.4255	.81865	57
4	.57453	.70194	1.4246	.81848	56
5	.57477	.70238	1.4237	.81832	**55**
6	.57501	.70281	1.4229	.81815	54
7	.57524	.70325	1.4220	.81798	53
8	.57548	.70368	1.4211	.81782	52
9	.57572	.70412	1.4202	.81765	51
10	.57596	.70455	1.4193	.81748	**50**
11	.57619	.70499	1.4185	.81731	49
12	.57643	.70542	1.4176	.81714	48
13	.57667	.70586	1.4167	.81698	47
14	.57691	.70629	1.4158	.81681	46
15	.57715	.70673	1.4150	.81664	**45**
16	.57738	.70717	1.4141	.81647	44
17	.57762	.70760	1.4132	.81631	43
18	.57786	.70804	1.4124	.81614	42
19	.57810	.70848	1.4115	.81597	41
20	.57833	.70891	1.4106	.81580	**40**
21	.57857	.70935	1.4097	.81563	39
22	.57881	.70979	1.4089	.81546	38
23	.57904	.71023	1.4080	.81530	37
24	.57928	.71066	1.4071	.81513	36
25	.57952	.71110	1.4063	.81496	**35**
26	.57976	.71154	1.4054	.81479	34
27	.57999	.71198	1.4045	.81462	33
28	.58023	.71242	1.4037	.81445	32
29	.58047	.71285	1.4028	.81428	31
30	.58070	.71329	1.4019	.81412	**30**
31	.58094	.71373	1.4011	.81395	29
32	.58118	.71417	1.4002	.81378	28
33	.58141	.71461	1.3994	.81361	27
34	.58165	.71505	1.3985	.81344	26
35	.58189	.71549	1.3976	.81327	**25**
36	.58212	.71593	1.3968	.81310	24
37	.58236	.71637	1.3959	.81293	23
38	.58260	.71681	1.3951	.81276	22
39	.58283	.71725	1.3942	.81259	21
40	.58307	.71769	1.3934	.81242	**20**
41	.58330	.71813	1.3925	.81225	19
42	.58354	.71857	1.3916	.81208	18
43	.58378	.71901	1.3908	.81191	17
44	.58401	.71946	1.3899	.81174	16
45	.58425	.71990	1.3891	.81157	**15**
46	.58449	.72034	1.3882	.81140	14
47	.58472	.72078	1.3874	.81123	13
48	.58496	.72122	1.3865	.81106	12
49	.58519	.72167	1.3857	.81089	11
50	.58543	.72211	1.3848	.81072	**10**
51	.58567	.72255	1.3840	.81055	9
52	.58590	.72299	1.3831	.81038	8
53	.58614	.72344	1.3823	.81021	7
54	.58637	.72388	1.3814	.81004	6
55	.58661	.72432	1.3806	.80987	**5**
56	.58684	.72477	1.3798	.80970	4
57	.58708	.72521	1.3789	.80953	3
58	.58731	.72565	1.3781	.80936	2
59	.58755	.72610	1.3772	.80919	1
60	.58779	.72654	1.3764	.80902	**0**
′	Cos	Cot	Tan	Sin	′

125° (305°) (234°) **54°**

NATURAL TRIGONOMETRIC FUNCTIONS

36° (216°) (323°) **143°**

′	Sin	Tan	Cot	Cos	′
0	.58779	.72654	1.3764	.80902	**60**
1	.58802	.72699	1.3755	.80885	59
2	.58826	.72743	1.3747	.80867	58
3	.58849	.72788	1.3739	.80850	57
4	.58873	.72832	1.3730	.80833	56
5	.58896	.72877	1.3722	.80816	**55**
6	.58920	.72921	1.3713	.80799	54
7	.58943	.72966	1.3705	.80782	53
8	.58967	.73010	1.3697	.80765	52
9	.58990	.73055	1.3688	.80748	51
10	.59014	.73100	1.3680	.80730	**50**
11	.59037	.73144	1.3672	.80713	49
12	.59061	.73189	1.3663	.80696	48
13	.59084	.73234	1.3655	.80679	47
14	.59108	.73278	1.3647	.80662	46
15	.59131	.73323	1.3638	.80644	**45**
16	.59154	.73368	1.3630	.80627	44
17	.59178	.73413	1.3622	.80610	43
18	.59201	.73457	1.3613	.80593	42
19	.59225	.73502	1.3605	.80576	41
20	.59248	.73547	1.3597	.80558	**40**
21	.59272	.73592	1.3588	.80541	39
22	.59295	.73637	1.3580	.80524	38
23	.59318	.73681	1.3572	.80507	37
24	.59342	.73726	1.3564	.80489	36
25	.59365	.73771	1.3555	.80472	**35**
26	.59389	.73816	1.3547	.80455	34
27	.59412	.73861	1.3539	.80438	33
28	.59436	.73906	1.3531	.80420	32
29	.59459	.73951	1.3522	.80403	31
30	.59482	.73996	1.3514	.80386	**30**
31	.59506	.74041	1.3506	.80368	29
32	.59529	.74086	1.3498	.80351	28
33	.59552	.74131	1.3490	.80334	27
34	.59576	.74176	1.3481	.80316	26
35	.59599	.74221	1.3473	.80299	**25**
36	.59622	.74267	1.3465	.80282	24
37	.59646	.74312	1.3457	.80264	23
38	.59669	.74357	1.3449	.80247	22
39	.59693	.74402	1.3440	.80230	21
40	.59716	.74447	1.3432	.80212	**20**
41	.59739	.74492	1.3424	.80195	19
42	.59763	.74538	1.3416	.80178	18
43	.59786	.74583	1.3408	.80160	17
44	.59809	.74628	1.3400	.80143	16
45	.59832	.74674	1.3392	.80125	**15**
46	.59856	.74719	1.3384	.80108	14
47	.59879	.74764	1.3375	.80091	13
48	.59902	.74810	1.3367	.80073	12
49	.59926	.74855	1.3359	.80056	11
50	.59949	.74900	1.3351	.80038	**10**
51	.59972	.74946	1.3343	.80021	9
52	.59995	.74991	1.3335	.80003	8
53	.60019	.75037	1.3327	.79986	7
54	.60042	.75082	1.3319	.79968	6
55	.60065	.75128	1.3311	.79951	**5**
56	.60089	.75173	1.3303	.79934	4
57	.60112	.75219	1.3295	.79916	3
58	.60135	.75264	1.3287	.79899	2
59	.60158	.75310	1.3278	.79881	1
60	.60182	.75355	1.3270	.79864	**0**
′	Cos	Cot	Tan	Sin	′

126° (306°) (233°) **53°**

37° (217°) (322°) **142°**

′	Sin	Tan	Cot	Cos	′
0	.60182	.75355	1.3270	.79864	**60**
1	.60205	.75401	1.3262	.79846	59
2	.60228	.75447	1.3254	.79829	58
3	.60251	.75492	1.3246	.79811	57
4	.60274	.75538	1.3238	.79793	56
5	.60298	.75584	1.3230	.79776	**55**
6	.60321	.75629	1.3222	.79758	54
7	.60344	.75675	1.3214	.79741	53
8	.60367	.75721	1.3206	.79723	52
9	.60390	.75767	1.3198	.79706	51
10	.60414	.75812	1.3190	.79688	**50**
11	.60437	.75858	1.3182	.79671	49
12	.60460	.75904	1.3175	.79653	48
13	.60483	.75950	1.3167	.79635	47
14	.60506	.75996	1.3159	.79618	46
15	.60529	.76042	1.3151	.79600	**45**
16	.60553	.76088	1.3143	.79583	44
17	.60576	.76134	1.3135	.79565	43
18	.60599	.76180	1.3127	.79547	42
19	.60622	.76226	1.3119	.79530	41
20	.60645	.76272	1.3111	.79512	**40**
21	.60668	.76318	1.3103	.79494	39
22	.60691	.76364	1.3095	.79477	38
23	.60714	.76410	1.3087	.79459	37
24	.60738	.76456	1.3079	.79441	36
25	.60761	.76502	1.3072	.79424	**35**
26	.60784	.76548	1.3064	.79406	34
27	.60807	.76594	1.3056	.79388	33
28	.60830	.76640	1.3048	.79371	32
29	.60853	.76686	1.3040	.79353	31
30	.60876	.76733	1.3032	.79335	**30**
31	.60899	.76779	1.3024	.79318	29
32	.60922	.76825	1.3017	.79300	28
33	.60945	.76871	1.3009	.79282	27
34	.60968	.76918	1.3001	.79264	26
35	.60991	.76964	1.2993	.79247	**25**
36	.61015	.77010	1.2985	.79229	24
37	.61038	.77057	1.2977	.79211	23
38	.61061	.77103	1.2970	.79193	22
39	.61084	.77149	1.2962	.79176	21
40	.61107	.77196	1.2954	.79158	**20**
41	.61130	.77242	1.2946	.79140	19
42	.61153	.77289	1.2938	.79122	18
43	.61176	.77335	1.2931	.79105	17
44	.61199	.77382	1.2923	.79087	16
45	.61222	.77428	1.2915	.79069	**15**
46	.61245	.77475	1.2907	.79051	14
47	.61268	.77521	1.2900	.79033	13
48	.61291	.77568	1.2892	.79016	12
49	.61314	.77615	1.2884	.78998	11
50	.61337	.77661	1.2876	.78980	**10**
51	.61360	.77708	1.2869	.78962	9
52	.61383	.77754	1.2861	.78944	8
53	.61406	.77801	1.2853	.78926	7
54	.61429	.77848	1.2846	.78908	6
55	.61451	.77895	1.2838	.78891	**5**
56	.61474	.77941	1.2830	.78873	4
57	.61497	.77988	1.2822	.78855	3
58	.61520	.78035	1.2815	.78837	2
59	.61543	.78082	1.2807	.78819	1
60	.61566	.78129	1.2799	.78801	**0**
′	Cos	Cot	Tan	Sin	′

127° (307°) (232°) **52°**

NATURAL TRIGONOMETRIC FUNCTIONS

38° (218°) (321°) **141°**

′	Sin	Tan	Cot	Cos	′
0	.61566	.78129	1.2799	.78801	**60**
1	.61589	.78175	1.2792	.78783	59
2	.61612	.78222	1.2784	.78765	58
3	.61635	.78269	1.2776	.78747	57
4	.61658	.78316	1.2769	.78729	56
5	.61681	.78363	1.2761	.78711	**55**
6	.61704	.78410	1.2753	.78694	54
7	.61726	.78457	1.2746	.78676	53
8	.61749	.78504	1.2738	.78658	52
9	.61772	.78551	1.2731	.78640	51
10	.61795	.78598	1.2723	.78622	**50**
11	.61818	.78645	1.2715	.78604	49
12	.61841	.78692	1.2708	.78586	48
13	.61864	.78739	1.2700	.78568	47
14	.61887	.78786	1.2693	.78550	46
15	.61909	.78834	1.2685	.78532	**45**
16	.61932	.78881	1.2677	.78514	44
17	.61955	.78928	1.2670	.78496	43
18	.61978	.78975	1.2662	.78478	42
19	.62001	.79022	1.2655	.78460	41
20	.62024	.79070	1.2647	.78442	**40**
21	.62046	.79117	1.2640	.78424	39
22	.62069	.79164	1.2632	.78405	38
23	.62092	.79212	1.2624	.78387	37
24	.62115	.79259	1.2617	.78369	36
25	.62138	.79306	1.2609	.78351	**35**
26	.62160	.79354	1.2602	.78333	34
27	.62183	.79401	1.2594	.78315	33
28	.62206	.79449	1.2587	.78297	32
29	.62229	.79496	1.2579	.78279	31
30	.62251	.79544	1.2572	.78261	**30**
31	.62274	.79591	1.2564	.78243	29
32	.62297	.79639	1.2557	.78225	28
33	.62320	.79686	1.2549	.78206	27
34	.62342	.79734	1.2542	.78188	26
35	.62365	.79781	1.2534	.78170	**25**
36	.62388	.79829	1.2527	.78152	24
37	.62411	.79877	1.2519	.78134	23
38	.62433	.79924	1.2512	.78116	22
39	.62456	.79972	1.2504	.78098	21
40	.62479	.80020	1.2497	.78079	**20**
41	.62502	.80067	1.2489	.78061	19
42	.62524	.80115	1.2482	.78043	18
43	.62547	.80163	1.2475	.78025	17
44	.62570	.80211	1.2467	.78007	16
45	.62592	.80258	1.2460	.77988	**15**
46	.62615	.80306	1.2452	.77970	14
47	.62638	.80354	1.2445	.77952	13
48	.62660	.80402	1.2437	.77934	12
49	.62683	.80450	1.2430	.77916	11
50	.62706	.80498	1.2423	.77897	**10**
51	.62728	.80546	1.2415	.77879	9
52	.62751	.80594	1.2408	.77861	8
53	.62774	.80642	1.2401	.77843	7
54	.62796	.80690	1.2393	.77824	6
55	.62819	.80738	1.2386	.77806	**5**
56	.62842	.80786	1.2378	.77788	4
57	.62864	.80834	1.2371	.77769	3
58	.62887	.80882	1.2364	.77751	2
59	.62909	.80930	1.2356	.77733	1
60	.62932	.80978	1.2349	.77715	**0**
′	Cos	Cot	Tan	Sin	′

128° (308°) (231°) **51°**

39° (219°) (320°) **140°**

′	Sin	Tan	Cot	Cos	′
0	.62932	.80978	1.2349	.77715	**60**
1	.62955	.81027	1.2342	.77696	59
2	.62977	.81075	1.2334	.77678	58
3	.63000	.81123	1.2327	.77660	57
4	.63022	.81171	1.2320	.77641	56
5	.63045	.81220	1.2312	.77623	**55**
6	.63068	.81268	1.2305	.77605	54
7	.63090	.81316	1.2298	.77586	53
8	.63113	.81364	1.2290	.77568	52
9	.63135	.81413	1.2283	.77550	51
10	.63158	.81461	1.2276	.77531	**50**
11	.63180	.81510	1.2268	.77513	49
12	.63203	.81558	1.2261	.77494	48
13	.63225	.81606	1.2254	.77476	47
14	.63248	.81655	1.2247	.77458	46
15	.63271	.81703	1.2239	.77439	**45**
16	.63293	.81752	1.2232	.77421	44
17	.63316	.81800	1.2225	.77402	43
18	.63338	.81849	1.2218	.77384	42
19	.63361	.81898	1.2210	.77366	41
20	.63383	.81946	1.2203	.77347	**40**
21	.63406	.81995	1.2196	.77329	39
22	.63428	.82044	1.2189	.77310	38
23	.63451	.82092	1.2181	.77292	37
24	.63473	.82141	1.2174	.77273	36
25	.63496	.82190	1.2167	.77255	**35**
26	.63518	.82238	1.2160	.77236	34
27	.63540	.82287	1.2153	.77218	33
28	.63563	.82336	1.2145	.77199	32
29	.63585	.82385	1.2138	.77181	31
30	.63608	.82434	1.2131	.77162	**30**
31	.63630	.82483	1.2124	.77144	29
32	.63653	.82531	1.2117	.77125	28
33	.63675	.82580	1.2109	.77107	27
34	.63698	.82629	1.2102	.77088	26
35	.63720	.82678	1.2095	.77070	**25**
36	.63742	.82727	1.2088	.77051	24
37	.63765	.82776	1.2081	.77033	23
38	.63787	.82825	1.2074	.77014	22
39	.63810	.82874	1.2066	.76996	21
40	.63832	.82923	1.2059	.76977	**20**
41	.63854	.82972	1.2052	.76959	19
42	.63877	.83022	1.2045	.76940	18
43	.63899	.83071	1.2038	.76921	17
44	.63922	.83120	1.2031	.76903	16
45	.63944	.83169	1.2024	.76884	**15**
46	.63966	.83218	1.2017	.76866	14
47	.63989	.83268	1.2009	.76847	13
48	.64011	.83317	1.2002	.76828	12
49	.64033	.83366	1.1995	.76810	11
50	.64056	.83415	1.1988	.76791	**10**
51	.64078	.83465	1.1981	.76772	9
52	.64100	.83514	1.1974	.76754	8
53	.64123	.83564	1.1967	.76735	7
54	.64145	.83613	1.1960	.76717	6
55	.64167	.83662	1.1953	.76698	**5**
56	.64190	.83712	1.1946	.76679	4
57	.64212	.83761	1.1939	.76661	3
58	.64234	.83811	1.1932	.76642	2
59	.64256	.83860	1.1925	.76623	1
60	.64279	.83910	1.1918	.76604	**0**
′	Cos	Cot	Tan	Sin	′

129° (309°) (230°) **50°**

NATURAL TRIGONOMETRIC FUNCTIONS

40° (220°) (319°) **139°**

′	Sin	Tan	Cot	Cos	′
0	.64279	.83910	1.1918	.76604	**60**
1	.64301	.83960	1.1910	.76586	59
2	.64323	.84009	1.1903	.76567	58
3	.64346	.84059	1.1896	.76548	57
4	.64368	.84108	1.1889	.76530	56
5	.64390	.84158	1.1882	.76511	**55**
6	.64412	.84208	1.1875	.76492	54
7	.64435	.84258	1.1868	.76473	53
8	.64457	.84307	1.1861	.76455	52
9	.64479	.84357	1.1854	.76436	51
10	.64501	.84407	1.1847	.76417	**50**
11	.64524	.84457	1.1840	.76398	49
12	.64546	.84507	1.1833	.76380	48
13	.64568	.84556	1.1826	.76361	47
14	.64590	.84606	1.1819	.76342	46
15	.64612	.84656	1.1812	.76323	**45**
16	.64635	.84706	1.1806	.76304	44
17	.64657	.84756	1.1799	.76286	43
18	.64679	.84806	1.1792	.76267	42
19	.64701	.84856	1.1785	.76248	41
20	.64723	.84906	1.1778	.76229	**40**
21	.64746	.84956	1.1771	.76210	39
22	.64768	.85006	1.1764	.76192	38
23	.64790	.85057	1.1757	.76173	37
24	.64812	.85107	1.1750	.76154	36
25	.64834	.85157	1.1743	.76135	**35**
26	.64856	.85207	1.1736	.76116	34
27	.64878	.85257	1.1729	.76097	33
28	.64901	.85308	1.1722	.76078	32
29	.64923	.85358	1.1715	.76059	31
30	.64945	.85408	1.1708	.76041	**30**
31	.64967	.85458	1.1702	.76022	29
32	.64989	.85509	1.1695	.76003	28
33	.65011	.85559	1.1688	.75984	27
34	.65033	.85609	1.1681	.75965	26
35	.65055	.85660	1.1674	.75946	**25**
36	.65077	.85710	1.1667	.75927	24
37	.65100	.85761	1.1660	.75908	23
38	.65122	.85811	1.1653	.75889	22
39	.65144	.85862	1.1647	.75870	21
40	.65166	.85912	1.1640	.75851	**20**
41	.65188	.85963	1.1633	.75832	19
42	.65210	.86014	1.1626	.75813	18
43	.65232	.86064	1.1619	.75794	17
44	.65254	.86115	1.1612	.75775	16
45	.65276	.86166	1.1606	.75756	**15**
46	.65298	.86216	1.1599	.75738	14
47	.65320	.86267	1.1592	.75719	13
48	.65342	.86318	1.1585	.75700	12
49	.65364	.86368	1.1578	.75680	11
50	.65386	.86419	1.1571	.75661	**10**
51	.65408	.86470	1.1565	.75642	9
52	.65430	.86521	1.1558	.75623	8
53	.65452	.86572	1.1551	.75604	7
54	.65474	.86623	1.1544	.75585	6
55	.65496	.86674	1.1538	.75566	**5**
56	.65518	.86725	1.1531	.75547	4
57	.65540	.86776	1.1524	.75528	3
58	.65562	.86827	1.1517	.75509	2
59	.65584	.86878	1.1510	.75490	1
60	.65606	.86929	1.1504	.75471	**0**
′	Cos	Cot	Tan	Sin	′

130° (310°) (229°) **49°**

41° (221°) (318°) **138°**

′	Sin	Tan	Cot	Cos	′
0	.65606	.86929	1.1504	.75471	**60**
1	.65628	.86980	1.1497	.75452	59
2	.65650	.87031	1.1490	.75433	58
3	.65672	.87082	1.1483	.75414	57
4	.65694	.87133	1.1477	.75395	56
5	.65716	.87184	1.1470	.75375	**55**
6	.65738	.87236	1.1463	.75356	54
7	.65759	.87287	1.1456	.75337	53
8	.65781	.87338	1.1450	.75318	52
9	.65803	.87389	1.1443	.75299	51
10	.65825	.87441	1.1436	.75280	**50**
11	.65847	.87492	1.1430	.75261	49
12	.65869	.87543	1.1423	.75241	48
13	.65891	.87595	1.1416	.75222	47
14	.65913	.87646	1.1410	.75203	46
15	.65935	.87698	1.1403	.75184	**45**
16	.65956	.87749	1.1396	.75165	44
17	.65978	.87801	1.1389	.75146	43
18	.66000	.87852	1.1383	.75126	42
19	.66022	.87904	1.1376	.75107	41
20	.66044	.87955	1.1369	.75088	**40**
21	.66066	.88007	1.1363	.75069	39
22	.66088	.88059	1.1356	.75050	38
23	.66109	.88110	1.1349	.75030	37
24	.66131	.88162	1.1343	.75011	36
25	.66153	.88214	1.1336	.74992	**35**
26	.66175	.88265	1.1329	.74973	34
27	.66197	.88317	1.1323	.74953	33
28	.66218	.88369	1.1316	.74934	32
29	.66240	.88421	1.1310	.74915	31
30	.66262	.88473	1.1303	.74896	**30**
31	.66284	.88524	1.1296	.74876	29
32	.66306	.88576	1.1290	.74857	28
33	.66327	.88628	1.1283	.74838	27
34	.66349	.88680	1.1276	.74818	26
35	.66371	.88732	1.1270	.74799	**25**
36	.66393	.88784	1.1263	.74780	24
37	.66414	.88836	1.1257	.74760	23
38	.66436	.88888	1.1250	.74741	22
39	.66458	.88940	1.1243	.74722	21
40	.66480	.88992	1.1237	.74703	**20**
41	.66501	.89045	1.1230	.74683	19
42	.66523	.89097	1.1224	.74664	18
43	.66545	.89149	1.1217	.74644	17
44	.66566	.89201	1.1211	.74625	16
45	.66588	.89253	1.1204	.74606	**15**
46	.66610	.89306	1.1197	.74586	14
47	.66632	.89358	1.1191	.74567	13
48	.66653	.89410	1.1184	.74548	12
49	.66675	.89463	1.1178	.74528	11
50	.66697	.89515	1.1171	.74509	**10**
51	.66718	.89567	1.1165	.74489	9
52	.66740	.89620	1.1158	.74470	8
53	.66762	.89672	1.1152	.74451	7
54	.66783	.89725	1.1145	.74431	6
55	.66805	.89777	1.1139	.74412	**5**
56	.66827	.89830	1.1132	.74392	4
57	.66848	.89883	1.1126	.74373	3
58	.66870	.89935	1.1119	.74353	2
59	.66891	.89988	1.1113	.74334	1
60	.66913	.90040	1.1106	.74314	**0**
′	Cos	Cot	Tan	Sin	′

131° (311°) (228°) **48°**

NATURAL TRIGONOMETRIC FUNCTIONS

42° (222°) — (317°) **137°**

′	Sin	Tan	Cot	Cos	′
0	.66913	.90040	1.1106	.74314	**60**
1	.66935	.90093	1.1100	.74295	59
2	.66956	.90146	1.1093	.74276	58
3	.66978	.90199	1.1087	.74256	57
4	.66999	.90251	1.1080	.74237	56
5	.67021	.90304	1.1074	.74217	**55**
6	.67043	.90357	1.1067	.74198	54
7	.67064	.90410	1.1061	.74178	53
8	.67086	.90463	1.1054	.74159	52
9	.67107	.90516	1.1048	.74139	51
10	.67129	.90569	1.1041	.74120	**50**
11	.67151	.90621	1.1035	.74100	49
12	.67172	.90674	1.1028	.74080	48
13	.67194	.90727	1.1022	.74061	47
14	.67215	.90781	1.1016	.74041	46
15	.67237	.90834	1.1009	.74022	**45**
16	.67258	.90887	1.1003	.74002	44
17	.67280	.90940	1.0996	.73983	43
18	.67301	.90993	1.0990	.73963	42
19	.67323	.91046	1.0983	.73944	41
20	.67344	.91099	1.0977	.73924	**40**
21	.67366	.91153	1.0971	.73904	39
22	.67387	.91206	1.0964	.73885	38
23	.67409	.91259	1.0958	.73865	37
24	.67430	.91313	1.0951	.73846	36
25	.67452	.91366	1.0945	.73826	**35**
26	.67473	.91419	1.0939	.73806	34
27	.67495	.91473	1.0932	.73787	33
28	.67516	.91526	1.0926	.73767	32
29	.67538	.91580	1.0919	.73747	31
30	.67559	.91633	1.0913	.73728	**30**
31	.67580	.91687	1.0907	.73708	29
32	.67602	.91740	1.0900	.73688	28
33	.67623	.91794	1.0894	.73669	27
34	.67645	.91847	1.0888	.73649	26
35	.67666	.91901	1.0881	.73629	**25**
36	.67688	.91955	1.0875	.73610	24
37	.67709	.92008	1.0869	.73590	23
38	.67730	.92062	1.0862	.73570	22
39	.67752	.92116	1.0856	.73551	21
40	.67773	.92170	1.0850	.73531	**20**
41	.67795	.92224	1.0843	.73511	19
42	.67816	.92277	1.0837	.73491	18
43	.67837	.92331	1.0831	.73472	17
44	.67859	.92385	1.0824	.73452	16
45	.67880	.92439	1.0818	.73432	**15**
46	.67901	.92493	1.0812	.73413	14
47	.67923	.92547	1.0805	.73393	13
48	.67944	.92601	1.0799	.73373	12
49	.67965	.92655	1.0793	.73353	11
50	.67987	.92709	1.0786	.73333	**10**
51	.68008	.92763	1.0780	.73314	9
52	.68029	.92817	1.0774	.73294	8
53	.68051	.92872	1.0768	.73274	7
54	.68072	.92926	1.0761	.73254	6
55	.68093	.92980	1.0755	.73234	**5**
56	.68115	.93034	1.0749	.73215	4
57	.68136	.93088	1.0742	.73195	3
58	.68157	.93143	1.0736	.73175	2
59	.68179	.93197	1.0730	.73155	1
60	.68200	.93252	1.0724	.73135	**0**
′	Cos	Cot	Tan	Sin	′

132° (312°) — (227°) **47°**

43° (223°) — (316°) **136°**

′	Sin	Tan	Cot	Cos	′
0	.68200	.93252	1.0724	.73135	**60**
1	.68221	.93306	1.0717	.73116	59
2	.68242	.93360	1.0711	.73096	58
3	.68264	.93415	1.0705	.73076	57
4	.68285	.93469	1.0699	.73056	56
5	.68306	.93524	1.0692	.73036	**55**
6	.68327	.93578	1.0686	.73016	54
7	.68349	.93633	1.0680	.72996	53
8	.68370	.93688	1.0674	.72976	52
9	.68391	.93742	1.0668	.72957	51
10	.68412	.93797	1.0661	.72937	**50**
11	.68434	.93852	1.0655	.72917	49
12	.68455	.93906	1.0649	.72897	48
13	.68476	.93961	1.0643	.72877	47
14	.68497	.94016	1.0637	.72857	46
15	.68518	.94071	1.0630	.72837	**45**
16	.68539	.94125	1.0624	.72817	44
17	.68561	.94180	1.0618	.72797	43
18	.68582	.94235	1.0612	.72777	42
19	.68603	.94290	1.0606	.72757	41
20	.68624	.94345	1.0599	.72737	**40**
21	.68645	.94400	1.0593	.72717	39
22	.68666	.94455	1.0587	.72697	38
23	.68688	.94510	1.0581	.72677	37
24	.68709	.94565	1.0575	.72657	36
25	.68730	.94620	1.0569	.72637	**35**
26	.68751	.94676	1.0562	.72617	34
27	.68772	.94731	1.0556	.72597	33
28	.68793	.94786	1.0550	.72577	32
29	.68814	.94841	1.0544	.72557	31
30	.68835	.94896	1.0538	.72537	**30**
31	.68857	.94952	1.0532	.72517	29
32	.68878	.95007	1.0526	.72497	28
33	.68899	.95062	1.0519	.72477	27
34	.68920	.95118	1.0513	.72457	26
35	.68941	.95173	1.0507	.72437	**25**
36	.68962	.95229	1.0501	.72417	24
37	.68983	.95284	1.0495	.72397	23
38	.69004	.95340	1.0489	.72377	22
39	.69025	.95395	1.0483	.72357	21
40	.69046	.95451	1.0477	.72337	**20**
41	.69067	.95506	1.0470	.72317	19
42	.69088	.95562	1.0464	.72297	18
43	.69109	.95618	1.0458	.72277	17
44	.69130	.95673	1.0452	.72257	16
45	.69151	.95729	1.0446	.72236	**15**
46	.69172	.95785	1.0440	.72216	14
47	.69193	.95841	1.0434	.72196	13
48	.69214	.95897	1.0428	.72176	12
49	.69235	.95952	1.0422	.72156	11
50	.69256	.96008	1.0416	.72136	**10**
51	.69277	.96064	1.0410	.72116	9
52	.69298	.96120	1.0404	.72095	8
53	.69319	.96176	1.0398	.72075	7
54	.69340	.96232	1.0392	.72055	6
55	.69361	.96288	1.0385	.72035	**5**
56	.69382	.96344	1.0379	.72015	4
57	.69403	.96400	1.0373	.71995	3
58	.69424	.96457	1.0367	.71974	2
59	.69445	.96513	1.0361	.71954	1
60	.69466	.96569	1.0355	.71934	**0**
′	Cos	Cot	Tan	Sin	′

133° (313°) — (226°) **46°**

44° (224°) (315°) **135°**

′	Sin	Tan	Cot	Cos	′
0	.69466	.96569	1.0355	.71934	**60**
1	.69487	.96625	1.0349	.71914	59
2	.69508	.96681	1.0343	.71894	58
3	.69529	.96738	1.0337	.71873	57
4	.69549	.96794	1.0331	.71853	56
5	.69570	.96850	1.0325	.71833	**55**
6	.69591	.96907	1.0319	.71813	54
7	.69612	.96963	1.0313	.71792	53
8	.69633	.97020	1.0307	.71772	52
9	.69654	.97076	1.0301	.71752	51
10	.69675	.97133	1.0295	.71732	**50**
11	.69696	.97189	1.0289	.71711	49
12	.69717	.97246	1.0283	.71691	48
13	.69737	.97302	1.0277	.71671	47
14	.69758	.97359	1.0271	.71650	46
15	.69779	.97416	1.0265	.71630	**45**
16	.69800	.97472	1.0259	.71610	44
17	.69821	.97529	1.0253	.71590	43
18	.69842	.97586	1.0247	.71569	42
19	.69862	.97643	1.0241	.71549	41
20	.69883	.97700	1.0235	.71529	**40**
21	.69904	.97756	1.0230	.71508	39
22	.69925	.97813	1.0224	.71488	38
23	.69946	.97870	1.0218	.71468	37
24	.69966	.97927	1.0212	.71447	36
25	.69987	.97984	1.0206	.71427	**35**
26	.70008	.98041	1.0200	.71407	34
27	.70029	.98098	1.0194	.71386	33
28	.70049	.98155	1.0188	.71366	32
29	.70070	.98213	1.0182	.71345	31
30	.70091	.98270	1.0176	.71325	**30**
31	.70112	.98327	1.0170	.71305	29
32	.70132	.98384	1.0164	.71284	28
33	.70153	.98441	1.0158	.71264	27
34	.70174	.98499	1.0152	.71243	26
35	.70195	.98556	1.0147	.71223	**25**
36	.70215	.98613	1.0141	.71203	24
37	.70236	.98671	1.0135	.71182	23
38	.70257	.98728	1.0129	.71162	22
39	.70277	.98786	1.0123	.71141	21
40	.70298	.98843	1.0117	.71121	**20**
41	.70319	.98901	1.0111	.71100	19
42	.70339	.98958	1.0105	.71080	18
43	.70360	.99016	1.0099	.71059	17
44	.70381	.99073	1.0094	.71039	16
45	.70401	.99131	1.0088	.71019	**15**
46	.70422	.99189	1.0082	.70998	14
47	.70443	.99247	1.0076	.70978	13
48	.70463	.99304	1.0070	.70957	12
49	.70484	.99362	1.0064	.70937	11
50	.70505	.99420	1.0058	.70916	**10**
51	.70525	.99478	1.0052	.70896	9
52	.70546	.99536	1.0047	.70875	8
53	.70567	.99594	1.0041	.70855	7
54	.70587	.99652	1.0035	.70834	6
55	.70608	.99710	1.0029	.70813	**5**
56	.70628	.99768	1.0023	.70793	4
57	.70649	.99826	1.0017	.70772	3
58	.70670	.99884	1.0012	.70752	2
59	.70690	.99942	1.0006	.70731	1
60	.70711	1.0000	1.0000	.70711	**0**
′	Cos	Cot	Tan	Sin	′

134° (314°) (225°) **45°**

Problem Answers

Chap. II, Section 5, pg. 21

1. 40 acres
   80 acres
   10 acres
   80 acres

3. Order Corner
   2 N¼ Cor. Sec. 36
   3 NW Sec. Cor. Sec. 25
   1 NE Sec. Cor. Sec. 35
   4 S¼ Cor. Sec. 26

5. Line tree described as being a certain species and size on line between two corners at a given distance from the starting corner.

   Bearing tree described as being a certain species and size at a given distance on a given bearing from a survey corner—useful in relocation of corner. Also known as witness tree.

Chap. III, Section 3, pg. 34

1. 1247.30 feet

3. 572.89

5. (a) short (b) 0.01 feet

7. (a) 198.57 feet (b) 23.85 feet

9. (a) 34.8% (b) 44.4 feet

11. 0.017 feet

13. 466.71 feet

15. (a) 8.75% (b) 112.0 feet

17. Slope = 38.2% S. D. = 236.69 feet

19. True length of tape is less than 100.00 feet by 0.08 feet.

21. 1097.40 feet; 893.98 feet.

23. 61.11 feet.

25. 14.11 links.

27. 98.59 feet, 101.43.

29. 100.05 feet.

Chap. IV, Section 3, pg. 52

1. 13°45′W

3. True Brg. AB = N37°W; True Brg. BC = N63°E;
True Brg. CD = S44°E; Azimuth CB = 243°00′;
Deflection ∡ @ C = 73°R; Azimuth DC = 316°

5. Correct declination = 20°45′E
Bearing = N89°15′W

7.

Station	Deflection ∡	Brg.
1+00	16°L	N 3°45′W
2+00	11°30′R	N 7°45′E
3+00	8°15′R	N16°00′E
4+00	4°30′L	N11°30′E
5+00	82°45′R	S85°45′E

9. Bearing BC = S2°53′E
Distance BC = 217.3 feet
Area ABC = 15,404 sq. feet

11. Bearing EF = N78°15′E

13. Mag. Decl. = 21°45′E

15. Mag. Brg. CD = S70°15′E
Azimuth CB = 225°45′
Brg. BA = N49°45′W
Mag. Brg. DE = S44°30′W

17. S3° 30′E.

19. 1800 degrees.

Chap. V, Section 5, pg. 71

1.			
Topographic	15.8(16)	35.8(36)	10.1(10)
Percent	24.2(24)	54.2(54)	15.6(15½)
Degree	13°30′	28°30′	8°45′

3. Slope distance = 82.4 feet

5. S.D. = 132.55 feet or 2.01 chains

7 S.D. = 95.4 links
S.D. = 132.6 links

9. D.E. = 115.91
H.D. = 270.80

11. 29.5 topog; 50.58 links.

13. 4.94 feet.

15. 100.55 links

Chap. VI, Section 4, pg. 92

1. Sta. 4.39

3. H.D. = 23.2 feet = 35.1 links
H.D. = 27.8 feet = 42.1 links
H.D. = 32.7 feet = 49.5 links

5. Correction is made by <u>subtracting</u> 1.42 feet

7.

STA.	H.D.	Abney	D.E.	Elev.
7.40				
			+16	
6.00		11.4		832
			-30	
	1.00 ch		-18	
		-18		880
			+40	

9. Topographic Abney = 52.6

11. 47 topog.

13. cotangent.

15. 22.65 topog; 34.3%.

Sample note forms

5 Sided Traverse

STA HD	FS	BS	INT ∡	CORRECT INT ∡	CALC BRG
C					
453.28'	N1°00'W	S0°30'E			N1°00'W
E					
430.82'	S32°30'E	N33°30'W			N32°30'W
B					
A			124°30'	124°30'	
288.22'	N62°00'W	S62°00'E			N62°00'W
E			95°00'	95°00'	
337.58	S31°00'W	N33°00'E			S33°00'W
D			101°00'	101°30'	
280.62'	S45°00'E	N48°00'W			S45°30'E
C			124°00'	124°00'	
287.46'	N79°30'E	S79°00'W			N79°30'E
B			93°30'	94°00'	
305.08	N6°30'W	S7°00'E			N6°30'W
A					

Σ = 539°00'

NE¼ SW¼, SEC. 18,
T11S, R5W, WM

FS_{BC} CHANGED TO N79°00'E APRIL 11, 1974

HYDE CP

BS_{DC} CHANGED TO N47°30'W WIDE COMP

THINN HC

ANSUM RC

EQUIPMENT

STAFF COMP #10

100' TAPE

2 PLUMB BOBS

1 HAND LEVEL

WEATHER

COOL & RAIN

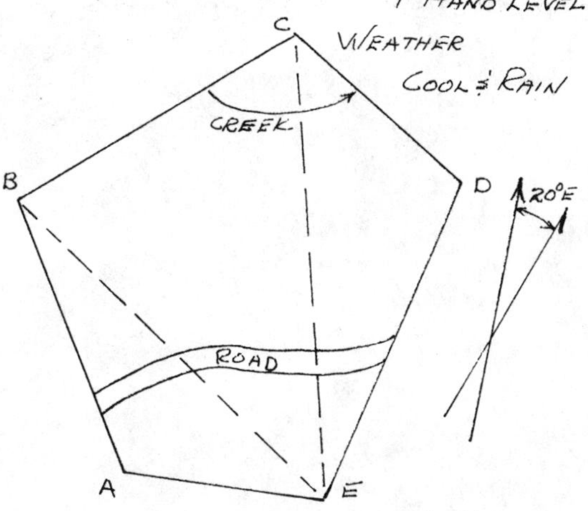

ANEROID BAROMETER

STA	OBS ELEV	TIME	ELEV CORRECTION	CORRECT ELEV
PEAVY HALL	275.0'	10:00AM	0	275'
LEWISBURG	300.0'	10:30AM	-10.4'	290'
LEWISBURG SADDLE	1000.0'	10:45AM	-15.6'	984'
CENTER SEC.4	800.0'	11:00AM	-20.8'	779'
SULPHUR SPRINGS	550.0'	2:00PM	-83.3'	467'
PEAVY HALL	400.0'	4:00PM	-125.0'	275'

APRIL 13, 1974

HYDE CP
JONES
MILLS
BOOM

ANEROID BAROMETER #17

WEATHER
COLD & WET

$$\text{CORRECTION} = \frac{400-275}{6\ \text{HOURS}} = \frac{125'}{6\ \text{HOURS}} = 20.83'/\text{HOUR}$$

MAP CONTROLS

STA	HD	TOPOG CLINOMETER	DE	ELEV	BRG
12.00				752'	
	1 CH		+6		
11.00		+6		746'	
	1 CH		+14		
10.00		+14		732'	
	1 CH		−10		
9.00		−10		742'	
	1 CH		−23		
8.00		−23		765'	
	2 CH		−14		
6.00		−7		779'	N86°00'E
6.00				779	
	1 CH		+11		
5.00		+11		768'	
	2 CH		+18		
3.00		+9		750'	
	2 CH		+13		
1.00		+6½		737'	
	1 CH		+4		
0.00		+4		733'	N4°00'W

12.00
FENCE 11.30
CREEK 9.72
21°E
6.00

6.00
21°E
1.54
ROAD
0.65
0.00

SEC. 18, T11S, R5W, WM
APRIL 20, 1973

J LOOK CP
M OVER
W HEER

EQUIPMENT
STAFF COMPASS #6
1-2CH TOPOG TAPE
2 CLINOMETERS #3 #11
1 AXE

WEATHER
CLOUDY & COOL

FROM STATION 0.00 TO STAKE: 1.65 CHAINS N10°E
STAKE IS 20.00 CHAINS NORTH & 20.00 CHAINS WEST OF E1/4 COR., SEC 18 T11S, R5W, WM.

DIFFENTIAL LEVELING

STA.	BS	HI	FS	ELEV
BM_1	12.5	763.0		750.5
TP	9.2	763.6	8.6	754.4
TP	11.5	768.7	6.4	757.2
TP	8.9	772.8	4.8	763.9
TBM_1	9.7	776.4	6.1	766.7
TP	7.9	780.0	4.3	772.1
TP	7.2	780.7	6.5	773.5
TP	10.6	784.7	6.6	774.1
TP	9.3	789.9	4.1	780.6
TBM_2	10.1	793.4	6.6	783.3
TP	6.6	792.8	7.2	786.2
TP	7.7	797.3	3.2	789.6
TP	4.3	794.9	6.7	790.6
TP	5.5	793.6	6.8	788.1
TP	6.3	790.7	9.2	784.4
TP	3.8	785.4	9.1	781.6
TP	1.3	778.9	7.8	777.6
TBM_3			6.3	772.6

$\Sigma = 132.4$ $\qquad\qquad \Sigma = 110.3$

$$\begin{array}{r} 772.6 \\ -\ 750.5 \\ \hline 22.1 \end{array}$$

$132.4 - 110.3 = 22.1$

BAKER CREEK
MAY 17, 1974

BRASS CAP - SOUTH END BRIDGE RAILING

SOFT CC
CLAY
MUUD

NAIL IN 30" FIR WEST SIDE ROAD

HAND LEVEL #7
LEVEL ROD #10

WEATHER

TOP OF ROCK EAST SIDE ROAD

CLEAR & WARM

IRON PIPE N 1/4 CORNER SEC. 5, T11N, R5E, WM

DEFLECTION ANGLE TRAVERSE

STA S.D.	H.O. VERT∡	ELEV D.E.	BS	FS	CALC BRG DEFL ∡
H		909.3'			
176.5'	176.5'	+3.5'		S8°30'W	S8°30'W
G	+2%	905.8'			5°30'R
198.6'	198.20'	+11.9	N3°00'E	S2°00'W	S3°00'W
F	+6%	893.9'			6°30'L
173.5'	173.3'	+8.7'	N8°30'E	S9°45'W	S9°30'W
E	+5%	885.2'			6°45'R
170.1'	169.9'	+8.5'	N3°00'E	S4°00'W	S2°45'W
D	+5%	876.7'			1°45'R
200.0'	200.00'	0	N2°15'E	S2°30'W	S1°00'W
C	0%	876.7'			5°30'R
182.7'	179.50'	-34.1'	N3°W	S3°00'E	S4°30'E
B	-19%	910.8'			81°30'R
165.3'	149.00'	+71.5'	N84°30'W	S85°30'E	S86°00'E
A	+48%	839.3'			4°45'R
			S89°45'W		N89°15'E

N½, SEC. 4, T11S, R5W, WM
JUNE 2, 1974

ROCK CC
HARD
PYLE

EQUIPMENT
1-STAFF COMPASS
1-200' ENGINEERS TAPE
2- % CLINOMETERS
2- PLUMB BOBS

WEATHER
CLOUDY & COOL

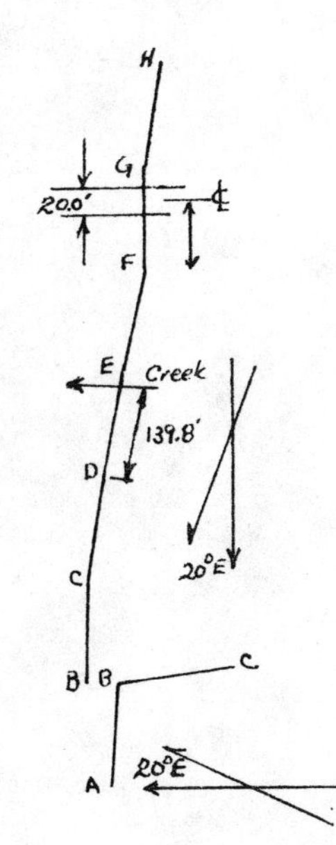